Microscopic Magnetic Resonance Imaging

Microscopic Magnetic Resonance Imaging

A Practical Perspective

Luisa Ciobanu

PAN STANFORD PUBLISHING

Published by

Pan Stanford Publishing Pte. Ltd.
Penthouse Level, Suntec Tower 3
8 Temasek Boulevard
Singapore 038988

Email: editorial@panstanford.com
Web: www.panstanford.com

British Library Cataloguing-in-Publication Data
A catalogue record for this book is available from the British Library.

Microscopic Magnetic Resonance Imaging: A Practical Perspective

Copyright © 2017 Pan Stanford Publishing Pte. Ltd.

Cover: Two neurons of *Aplysia californica* imaged with MRM.

For photocopying of material in this volume, please pay a copying fee through the Copyright Clearance Center, Inc., 222 Rosewood Drive, Danvers, MA 01923, USA. In this case permission to photocopy is not required from the publisher.

ISBN 978-981-4774-71-0 (Paperback)
ISBN 978-981-4774-42-0 (Hardback)
ISBN 978-1-315-10732-5 (eBook)

Printed in Great Britain by Ashford Colour Press Ltd.

To my parents:

Nature or nurture, I owe it all to you.

Contents

Section III

APPLICATIONS

SECTION **IV**

CONCLUSION

Preface

This book would not have existed without the support from my family and colleagues. When I received the invitation to write it for Pan Stanford Publishing, my husband, Catalin, was the first to persuade me to accept the proposal. He helped me start and finish the book; he had the patience to proofread it entirely before submission! At NeuroSpin, I have received constant encouragement from many people, in particular, from Drs. Denis Le Bihan and Cyril Poupon.

I want to thank all the members of the NeuroPhysics team: Tangi, Khieu, Yoshi, Pavel, Tom, and Gabrielle. Some of them contributed with ideas, others with figures or images, and overall everyone was extremely helpful and supportive. I would also like to thank Drs. Andrew Webb, Romuald Nargeot, Jing-Rebecca Li, and Tangi Roussel for reading the early manuscript versions and providing corrections on specific chapters.

I am grateful to have been able to include images acquired on the unique 17.2 T imaging system at NeuroSpin. These acquisitions were possible with the support received from CEA-Saclay and from the French National Agency of Research (ANR), funder of my research.

Writing this book took time; I thank my family and friends, and especially Robert, my son, for their understanding and patience in seeing "less of me" while I was working on it.

Luisa Ciobanu
Paris, 2017

SECTION I

INTRODUCTION

Chapter 1

About This Book

This book aims to provide a simple introduction to magnetic resonance microscopy (MRM) emphasizing practical aspects relevant to high magnetic fields. The text is intended for the beginners in the field of MRM or for those planning to incorporate high-resolution magnetic resonance imaging (MRI) in their neuroscience studies. For a more advanced level, we recommend *Principles of Magnetic Resonance Microscopy*, by P. Callaghan (Callaghan, 1991).

The first chapters are mainly pedagogical, introducing the reader to the hardware (Chapter 2), image acquisition principles (Chapter 3), various pulse sequences (Chapter 4), contrast mechanisms and image artifacts (Chapter 5), and specifics of sample preparation for microscopy studies (Chapter 6). As we move from the generic aspects of MRM to MRM applications, readers will notice a change in the presentation approach. Specifically, the following three chapters (Chapters 7–9) are written in the form of reviews. Chapter 7 surveys the most relevant experimental developments over the past three decades. In view of biological applications, it also introduces the *Aplysia californica*, one of the most used model systems in MRM studies with single-neuron resolution. Chapters 8 and 9 focus on two specific applications: high-resolution diffusion and functional studies. We note here that these applications constitute just a

Microscopic Magnetic Resonance Imaging: A Practical Perspective
Luisa Ciobanu
Copyright © 2017 Pan Stanford Publishing Pte. Ltd.
ISBN 978-981-4774-71-0 (Paperback), 978-981-4774-42-0 (Hardback), 978-1-315-10732-5 (eBook)
www.panstanford.com

small part of the full panel of possible MRI investigations at the microscopic scale. The choice was dictated primarily by our own expertise in the field. We also caution the reader that we are restricting our discussion to the use of MRM for studying biological systems with particular focus on the nervous system. The utility of MRM is of course much broader and includes material and chemical sciences, microfluidics, food industry, and plant physiology. For those interested in these other types of applications, we suggest an excellent monograph edited by Sarah Codd and Joseph Seymour (Codd and Seymour, 2009). Finally, in the last chapter (Chapter 10), we discuss some of the most probable future directions of MRM.

The majority of images included have been acquired specifically for this book; the corresponding experimental parameters are listed in the Appendix.

SECTION II

BASICS

Chapter 2

Hardware

While the same hardware elements are necessary for performing conventional MRI and MRM, there are certain technical demands specific to microscopy. This chapter briefly introduces the reader to the main components of an MRI scanner and to the technical challenges imposed by high-resolution MR microscopy. An important part is dedicated to the design and construction of radiofrequency coils dedicated to MRM.

2.1 The Main Magnet

The purpose of the main magnet is to generate a strong, uniform, static magnetic field, known as the B_0 field, in order to polarize the nuclear spins in the object being imaged. This polarization leads to a net magnetization which is proportional to the spin density of the object and to the strength of the B_0 field.[a] In the image, the signal level relative to noise, typically expressed as the average signal divided by the standard deviation of the noise and

---

[a]In the International System of Units, the strength of magnetic field is measured in Tesla (T).

Microscopic Magnetic Resonance Imaging: A Practical Perspective
Luisa Ciobanu
Copyright © 2017 Pan Stanford Publishing Pte. Ltd.
ISBN 978-981-4774-71-0 (Paperback), 978-981-4774-42-0 (Hardback), 978-1-315-10732-5 (eBook)
www.panstanford.com

referred to as signal-to-noise-ratio (SNR)[b] is proportional to the net magnetization. In MR microscopy one aims to distinguish fine spatial features requiring high spatial resolutions. This implies that the voxel volume is several orders of magnitude smaller than the typical volume resolution obtained in clinical settings. The small voxel size reduces the number of spins generating the MR signal. The only way to increase the available net magnetization is to use high magnetic fields. Typically, high-resolution MRM experiments are performed at magnetic fields higher than 7 T and as high as 21 T. Another characteristic of the main magnet is its homogeneity, expressed in parts per million (ppm), over a spherical volume with a certain diameter. Generally a homogeneity of 10–50 ppm is acceptable. For MRM, this requirement is easily achievable as the objects to be imaged are usually small (several millimeters). However, other elements can deteriorate the B_0 homogeneity as we will see later in the book.

2.2 Radiofrequency Coils

The radiofrequency coils (RF) are used to excite the spin system (transmitters) and to detect the MR signal (receivers). The transmitter generates a rotating magnetic field, known as the B_1 field, perpendicular to B_0. This B_1 field rotates the magnetization, M_0, initially aligned with B_0, about its axis (Fig. 2.1). The pulse of energy used to generate this rotation is called *RF pulse*. The angle of rotation of the magnetization, α in Fig. 2.1, is called the *tip* or *flip* angle and it depends on the length and the amplitude of the pulse. When the RF pulse is turned off, the transverse component of the magnetization precesses about the main magnetic field at a frequency, known as the Larmor frequency (ω_0), determined by the nucleus under study and the strength of the main magnetic field ($\omega_0 = \gamma B_0$, where γ is the gyromagnetic ratio[c]).

---

[b] In this chapter, we will use the term SNR as an overall measure of the detection sensitivity. A detailed description of its dependence on the hardware and imaging parameters will be provided in Chapter 3.

[c] $\gamma_{proton} = 42.58$ MHz/T.

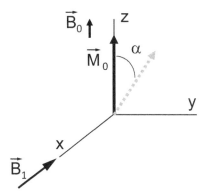

Figure 2.1 The B_1 field rotates the magnetization, M_0, toward the transverse plane.

The receiver coil converts the precessing magnetization into an electrical voltage which constitutes the MR signal. Both the transmitter and the receiver circuitry resonate at the Larmor frequency. It is desirable that the transmit coil produces a very uniform B_1 field and that the receiver coil has a high sensitivity (dictated by the smallest signal that can be detected: the higher the sensitivity the smaller the minimum detected signals). In many cases, and especially in MR microscopy, a single coil is used as the transmitter and the receiver and is called *transceiver*.

2.2.1 Basic Coil Designs

Depending on the extent of the region they cover there are two main types of RF coils: volume and surface. Volume coils surround the object to be imaged while surface coils are placed adjacent to it. Regardless of their type, one can show (as we detail in the next section) that for optimum detection, the size and geometry of the RF coil should closely match the sample shape. Therefore, MR microscopy imposes the use of small coils: *microcoils*. As a matter of definition, there is no general consensus regarding the size of a microcoil; in this book, we will call "microcoil" any RF coil smaller than 5 mm.

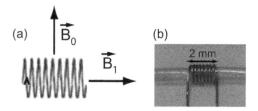

Figure 2.2 (a) Schematic of a solenoidal RF coil showing the direction of the B_1 field with respect to B_0. (b) Photograph of a manually wound microcoil. The coil was wrapped on a 600 μm diameter polyimide tubing using 100 μm diameter copper wire.

2.2.1.1 Solenoidal coils

Among the many volume coil designs (saddle, birdcage, etc.) the most widespread in MR microscopy experiments is the solenoidal geometry (Fig. 2.2). This choice is justified by the high uniformity of the generated B_1 field. In addition, even in miniaturized forms, these coils are relatively easy to fabricate by winding a thin wire on small diameter capillaries (Webb, 2010). The miniaturization of solenoidal coils is limited only by the wire diameter. Manually wound microcoils with diameters as small as 60 μm have been reported using copper wire with 10 μm diameter (Ciobanu, 2002).

While solenoids are relatively easy to build there are several factors which must be taken into consideration in their design stage. The sensitivity of a solenoid, defined as the B_1 field produced by unit current i, can be expressed as:

$$\frac{B_1(y)}{i} = n\mu_0 \left[\frac{0.5 + \frac{y}{l_{\text{coil}}}}{\sqrt{d_{\text{coil}}^2 + (l_{\text{coil}} + 2y)^2}} + \frac{0.5 - \frac{y}{l_{\text{coil}}}}{\sqrt{d_{\text{coil}}^2 + (l_{\text{coil}} - 2y)^2}} \right],$$

$$(2.1)$$

where y is the distance from the center of the solenoid along its long axis, μ_0 is the magnetic permeability of vacuum, n is the number of turns, and d_{coil} and l_{coil} are the coil diameter and length, respectively. The deviation of the B_1 field at the edges of a given sample of length l_{sample} relative to the field in the center of the coil can be derived from Eq. 2.1:

$$\frac{B_1(0) - B_1(edge)}{B_1(0)} = 1 - \frac{1}{2}\sqrt{1 + \left(\frac{d_{coil}}{l_{coil}}\right)^2}$$

$$\times \left[\frac{\frac{l_{coil}}{d_{coil}} + \frac{l_{sample}}{d_{coil}}}{\sqrt{1 + \left(\frac{l_{coil}}{d_{coil}} + \frac{l_{sample}}{d_{coil}}\right)^2}} + \frac{\frac{l_{coil}}{d_{coil}} - \frac{l_{sample}}{d_{coil}}}{\sqrt{1 + \left(\frac{l_{coil}}{d_{coil}} - \frac{l_{sample}}{d_{coil}}\right)^2}} \right] \quad (2.2)$$

Considering a spherical sample centered inside the solenoid and imposing a maximum deviation of the B_1 field of 20% one obtains an optimum coil length of approximately 1.5 times its diameter. In addition, in order to maximize its sensitivity, the coil diameter has to be the smallest possible for a given sample size. Once the diameter and length are fixed the other coil characteristics (number of turns, spacing and wire diameter) have to be appropriately chosen. According to Hoult and Richards (1976), the optimum spacing between turns is approximately 1.5 times the wire diameter. For non-conducting samples, Minard and Wind (2001a,b) showed that the SNR per unit sample volume is maximized for a coil made using thin wire and large number of turns. On the contrary, for conducting samples the coils should have fewer turns and thicker wire. A more in-depth analysis of the sensitivity of solenoids operating at high magnetic fields is presented in Section 2.2.3.

As we will see in the later chapters, the quality of MR images is greatly affected by the homogeneity of the static magnetic field. The close proximity of the RF coil to the sample will distort the B_0 field and lead to severe image artifacts. The magnitude of these artifacts increases with the strength of the magnetic field, B_0. The easiest workaround is to immerse the coil in a material with magnetic susceptibility similar to that of the coil (Peck, 1995). In this way, one mimics an infinite cylinder of given susceptibility in which the static field is homogeneous. Images of a water phantom acquired at 17.2 T using a solenoidal RF coil immersed and not-immersed in a susceptibility matching fluid (Fluorinert-FC40 - 3 M, Minneapolis, MN) are shown in Fig. 2.3.

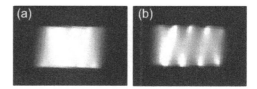

Figure 2.3 The impact on image quality of using Fluorinert-FC40 as a coil surrounding medium. (a) Water phantom image acquired with the FC-40 fluid present. (b) Water phantom image acquired in the absence of the fluid. Significant image inhomogeneity is observed in the latter case due to coil windings. Operating frequency 730 MHz. The acquisition parameters are listed in Appendix A.

2.2.1.2 Surface coils

Despite their advantageous properties, solenoids may not be the best design choice in certain experimental settings. Such examples include situations in which the solenoidal geometry is not compatible with the geometry of the sample or when easy sample access during experimentation is needed. In these cases, surface coils are convenient alternatives. The simplest surface coil design consists of a single circular loop of wire (Fig. 2.4).

When compared to the solenoidal coils described before, surface coils provide very high localized SNR, which, however, decreases rapidly with increasing distance from the coil plane. According to the Biot–Savart law (Jackson, 1975), the sensitivity of a circularly

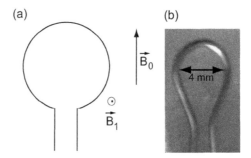

Figure 2.4 (a) Schematic of a single loop RF surface coil showing the direction of the B_1 field with respect to B_0. (b) Photograph of a single loop coil fabricated manually.

shaped coil is given by the expression:

$$\frac{B_1(z)}{i} = \frac{\mu_0}{2R_{\text{coil}}\left(1 + \frac{z^2}{R_{\text{coil}}^2}\right)^{\frac{3}{2}}},$$ (2.3)

where R_{coil} is the coil radius and z is the distance from the coil.

From Eq. 2.3 one can clearly see that the penetration depth, defined as the distance where the B_1 field decreases to 37%[d] of its maximum, is determined by the radius of the coil.

Single loops with diameters larger than 1 mm can be easily built manually, while submillimeter loops or more sophisticated designs (spiral, butterfly) require the use of photolithography and microfabrication techniques. Spiral coils are often used in MR microscopy as they have higher sensitivity than single loops assuming the number of turns is smaller than an optimum number beyond which the resistive losses overcome their contribution to SNR gain (Eroglu, 2003).

Arrays of microcoils, consisting of several electrically isolated, coil elements have also been demonstrated experimentally; however, their construction is challenging due to the small size (Gruschke, 2012). The main advantage of using coil arrays is the reduced scanning time resulting from parallel image acquisitions.

2.2.2 RF Circuit Design

RF coils can be modeled as RLC circuits. Most coils can be approximated by the circuit represented in Fig. 2.5, consisting of an inductor (L) placed in series with a resistor (R) and in parallel with a capacitor (C). For optimum sensitivity the MRI probes must be properly *tuned* and *matched*. The tuning adjusts the resonance frequency of the circuit to the Larmor frequency ω_0 imposed by the external magnetic field. During transmission impedance-matching ensures the optimum power transfer from the RF amplifier (50 Ω output impedance) to the RF coil. Improper matching will require large amounts of power to generate the desired pulses, possibly leading to electrical arcing (unwanted electrical discharge) of the coil. During reception impedance-matching provides efficient power

---

[d] $1/e = 1/2.718 = 0.368$; where e is the base of the natural logarithm.

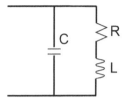

Figure 2.5 RLC modelization of an RF coil.

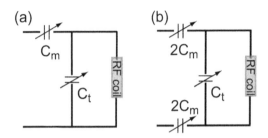

Figure 2.6 (a) Standard RF circuit. (b) Balanced RF circuit.

transfer from the coil to the signal preamplifier, ensuring high sensitivity and good SNR. The simplest scheme used to match the coil to a 50 Ω impedance is represented in Fig. 2.6a, in which C_t and C_m are variable tuning and matching capacitors. A slightly modified design, called a *balanced* circuit, is often used in order to minimize the noise introduced by conductive samples. In this circuit two matching capacitors, with capacitances approximately twice the matching capacitance used in the standard design, are placed on each side of the RF coil (Fig. 2.6b). The tuning/matching circuit is usually implemented on printed circuit boards (PCB). Typical values for variable capacitors range from 0.5 to 15 pF.

The highest frequency to which a coil can be impedance-matched is given by its self-resonant frequency:

$$\omega_{self} = \sqrt{\frac{1}{LC} - \frac{R^2}{L^2}} \tag{2.4}$$

For a solenoid the inductance (L) and capacitance (C) can be calculated according to the following empirical formulae (Fukushima,

1981; Medhurst, 1947):

$$L = \frac{n^2 d_{\text{coil}}^2}{46 d_{\text{coil}} + 100 l_{\text{coil}}} \tag{2.5}$$

$$C = d_{\text{coil}} \left(0.1126 \frac{l_{\text{coil}}}{d_{\text{coil}}} + 0.08 + 0.27 \sqrt{\frac{d_{\text{coil}}}{l_{\text{coil}}}} \right), \tag{2.6}$$

where n is the number of turns, and d_{coil} and l_{coil} are the solenoid's diameter and length expressed in centimeters. In Eqs. 2.5 and 2.6, L and C are expressed in microhenries and picofarads, respectively. At high operating frequency the estimation of coil resistance requires taking into consideration several additional factors, including *skin depth effects* and *proximity effects*. These effects will be discussed in detail in Section 2.2.3. The theoretical self-resonance frequency of solenoidal microcoils is in the gigahertz regime, with smaller diameter coils resonating at higher frequencies. In practice, there are several factors which can influence the self-resonance frequency of the coil. The leads of the coil increase its inductance thereby reducing its self-resonance frequency. A conductive sample introduces an extra capacitance, which also reduces the resonant frequency of the coil. The magnitude of these effects depends on the size of the microcoil. The impact of sample loading is lesser for smaller coils, while the lead inductance becomes more important as the coil length decreases.

2.2.3 Coil Performance

A standard measure used to characterize RF coils is the quality factor, Q, indicating the energy loss relative to the amount of energy stored within the system. A high Q value signifies a low rate of energy loss and therefore an efficient coil. In practice the easiest way to measure the Q-factor is through reflexion-type measurements (the same measurements used to verify the matching and tuning of the coil) and applying the following definition:

$$Q = \frac{\omega_0}{\Delta \omega_0}, \tag{2.7}$$

where ω_0 is the resonant frequency and $\Delta \omega_0$ is the bandwidth measured at half power (at -3 dB from the baseline). Considering the coil modeled by the RLC circuit in Fig. 2.5 the Q factor can be calculated as follows:

$$Q = \frac{1}{R}\sqrt{\frac{L}{C}} \qquad (2.8)$$

From Eq. 2.8, higher Q-factors are obtained for lower resistances, R. For a loaded coil R is the sum of the coil and sample resistances. For very small coils (<1 mm) the losses are mainly due to the coil, meaning that the loaded and unloaded Q-factors are very similar. As microcoils are constructed using thin wire (high resistance) their Q-value is smaller than that of coils used in preclinical and clinical imaging, with typical values under 100 for resonance frequencies between 400 and 750 MHz. Moreover, operating at very high frequencies further increases the resistance due to *skin depth effects*. The skin depth effect refers to the non-uniform distribution of an alternative current as it passes through a conductor, presenting a higher density at the surface (skin) of the conductor. The skin depth, δ, is defined as the depth from the surface of the conductor at which the current density decreases to $1/e$ of its value at the surface and can be calculated according to

$$\delta = \sqrt{\frac{2}{\omega_0 \mu \sigma}}, \qquad (2.9)$$

where σ and μ are the conductivity and the magnetic permeability of the wire, respectively, and ω_0 is the operating frequency. As the frequency increases, the effective cross-sectional area of the wire is reduced, leading to an increase in its resistance. At 730 MHz (17 T) the skin depth of copper is 2.44 μm. In the case of a closely wound solenoid the interaction between the magnetic fields within the different turns induces eddy currents which further restrict the regions (*proximity effects*) in which the current flows, increasing again the resistance. The effective coil resistance is often expressed as $R_{\text{eff}} = R_{\text{DC}}(1 + F + G)$, where F and G are the skin and proximity effect factors, respectively, and R_{DC} is the direct current (DC) resistance.

Theoretical calculations taking into consideration the two effects described above agree well with experimental results and show that the coil sensitivity is inversely proportional to the coil diameter for larger coils wound with thicker wires, while for smaller coils and thinner wires it changes with the square root of the coil diameter. While the transition point depends on the coil geometry and the operating frequency, as a rule of thumb, coils with diameters smaller than 100 μm fall into the second category (square-root variations) (Peck, 1995).

2.2.4 Other Types of Coils

2.2.4.1 Inductively coupled coils

The placement of tuning and matching capacitors, as well as of other electrical components (cables, for example), close to the RF microcoil leads to significant susceptibility artifacts. Moving these elements farther away requires long coil leads which introduce other deleterious effects (changes in coil characteristics and reduced SNR). An elegant way to overcome this problem is to use an *inductively coupled circuit*. In this design the microcoil forms a stand alone, self-resonant circuit which is not impedance matched and is not physically connected with the transmission or reception circuits. Instead, the microcoil is inductively coupled to a larger coil which is interfaced with the spectrometer (Fig. 2.7).

The coupling constant between the two coils, k, is defined as

$$k = \frac{M_{mb}}{\sqrt{L_m L_b}}, \quad 0 \leq k \leq 1, \tag{2.10}$$

where M_{mb} is the mutual inductance between the two coils and L_m and L_b are their respective self-inductances. Depending on the value of k the circuit can be weakly, critically or strongly coupled, with the

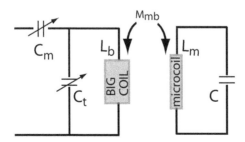

Figure 2.7 Schematic of an inductively coupled probe. M_{mb} is the mutual inductance between the two coils.

critical coupling value defined as

$$k_c = \frac{1}{\sqrt{Q_m Q_b}},$$

(2.11)

where Q_m and Q_b are the quality factors of the two coils. For optimum SNR k should be slightly larger than k_c (slightly over coupled regime).

Inductively coupled circuits employing different coil geometries have been reported. Most of them use custom designed microcoils (solenoids or surface loops) coupled with standard, commercially available resonators (Nabuurs, 2011; Tang and Jerschow, 2010). Given their small size, one can simultaneously use several microcoils inductively coupled to the same large coil for multiple sample imaging (Wang, 2008).

2.2.4.2 Cryogenically cooled coils

In cases in which the coil resistance is larger than that of the sample, the Q-factor, and therefore the SNR in an MR experiment, can be improved by reducing the coil resistance. This can be accomplished by cryocooling the coil, which can be either a standard copper coil or a high-temperature superconducting (HTS) coil.

Let us assume a copper coil cooled at 77 K (liquid nitrogen temperature). Using the temperature coefficient of copper $\alpha = 0.004$ K^{-1}, it follows that the coil resistance decreases by a factor of eight at 77 K compared to room temperature. Taking into consideration the skin effects discussed previously (a $\sqrt{8}$ correction) we arrive at a theoretical Q-factor improvement of approximately 2.8-fold.

Cryocooling the coil while maintaining the sample at room temperature is difficult for microcoils, as it implies significantly increasing the distance between the coil and the sample which leads to SNR deterioration. The development of micro-fluidic cooling devices appears to be a promising approach in terms of minimizing the coil-sample distance. Using this approach in combination with an inductively coupled surface spiral microcoil (2 mm inner diameter), Koo et al. (2014) achieved a factor of 2.6 improvement in the Q-factor.

2.3 Gradient Coils

In 1973 Paul Lauterbur demonstrated that one can generate MR images by superimposing a weaker, spatially variable magnetic field on the uniform, static field B_0. Because this magnetic field gradient, G, causes additional fields, much smaller than the B_0 field, the local Larmor frequency is position dependent:

$$\omega(\vec{r}) = \gamma B_0 + \gamma \vec{G} \cdot \vec{r} \qquad (2.12)$$

Different parts of the sample will therefore have different resonance frequencies depending on their location. Moreover, the strength of the MR signal at each frequency will be proportional to the number of spins at that frequency and thus at the corresponding position in space. An MR image is obtained by mapping the signal intensity throughout the sample.[e]

The additional, time-varying magnetic field necessary to encode frequency-position information is generated using gradient coils. The typical geometries of these coils are referred to as concentric Golay saddle (x and y coils) and Maxwell pairs (z coil) (Fig. 2.8). The performance of the gradient coils is specified by the maximum gradient strength, rise time and slew rate. The maximum gradient strength is expressed in units of T/m or, more commonly, mT/m. Clinical and preclinical scanners have gradient strengths of tens and hundreds of mT/m, while for MR microscopy much higher strengths (thousands of mT/m) are used. The rise time represents the time necessary to increase the gradient from zero to its maximum strength. The slew rate is defined as the maximum strength divided by the rise time. Higher and faster switching magnetic field gradients allow faster imaging and higher spatial resolutions.

Ideally imaging gradients should produce magnetic fields with intensities increasing linearly with distance from the center of the magnet (isocenter). In practice, gradient linearity is difficult to achieve over large regions as it falls off significantly as one moves away from isocenter. Fortunately, MR microscopy operates over small regions of interest, and therefore this is not a major concern.

The small size of the coils used in MRM facilitates the design and construction of extremely high performance gradient coils capable

---

[e]For an introduction to the basic imaging principles, see Chapter 3.

Golay coil

Maxwell pair coil

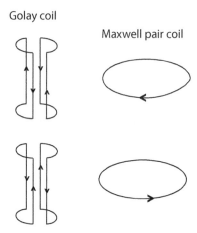

Figure 2.8 Schematic drawings of typical gradient coil geometries: Golay coils for G_x and G_y and Maxwell pair coil for G_z.

of very rapid switching and strong magnetic field gradients. As an example, Seeber et al. (2000) developed a triaxial gradient system, producing gradients greater than 15 T/m in all three directions with a very short 10 µs rise time. Using this gradient system and very small solenoidal microcoils Ciobanu et al. obtained images with resolutions as high as 3.5 µm (Ciobanu, 2002).

2.4 The Elusive "Key" Component

One question often asked when talking about high-resolution MR microscopy is: what is the key hardware component? The answer is complex.

The maximum theoretical spatial resolution which can be achieved in MR microscopy is dictated by the performance of the magnetic field gradients. In order to measure the location of a spin with a precision Δx, one must necessarily measure the frequency with a corresponding precision $\Delta \omega$. According to the uncertainty principle, this measurement requires a minimum acquisition time $T_{acq} \geq 1/\Delta \omega$. As a result, in agreement with Eq. 2.12, Δx should satisfy the relationship: $\Delta x \geq 1/(\gamma G T_{acq})$. Therefore, for high resolution, one requires that the product of the gradient strength and the time of the measurement, T_{acq}, be large. If, however, the spins diffuse spatially during the acquisition, then the spatial resolution will be further limited according to $\Delta x \geq \sqrt{2DT_{acq}}$, where D is the diffusion coefficient (approximately 2×10^{-5} cm^2/s for water at room temperature).[f] It is clear that, in order to achieve high spatial resolution, it is necessary to acquire data fast, minimizing T_{acq} in order to reduce the impact of diffusion while maintaining a large product $\gamma G T_{acq}$ for adequately resolving the frequency. The only way to achieve both goals is to use strong and fast switching magnetic field gradients. For example, imaging a biological tissue with 2 μm spatial resolution requires $T_{acq} \leq 2$ ms, using a diffusion coefficient of 1×10^{-5} cm^2/s. Combining this result with the uncertainty principle we find that magnetic field gradients greater than 6 T/m are needed. Moreover, having the gradients with the required specifications does not guarantee that the desired images can be obtained. Besides having the strong gradients capable of encoding high-resolution images is it equally important to be able to obtain an adequate SNR in order to render the images useful. We see then that there is no single "key" component: for best results one has to use very strong magnetic field gradients in combination with well-designed RF coils, and when possible, with high magnetic field systems.

---

[f]In this discussion we ignore the resolution limits imposed by relaxation times and susceptibility effects and focus only on molecular diffusion.

Chapter 3

Image Formation

In this chapter, we describe the evolution of the nuclear spin magnetization in the presence of an external magnetic field and under the influence of magnetic field gradients. We also introduce the $\vec{k}$-space and define basic image characteristics such as spatial resolution and signal-to-noise ratio.

3.1 The Bloch Equation

The behavior of the magnetization, $\vec{M}$, in the presence of an external magnetic field, $\vec{B}_1$ (generated by the RF coil), is described by the Bloch equation (Abragam, 1961):

$$\frac{d\vec{M}}{dt} = \gamma \vec{M} \times \vec{B}_{\text{eff}} - \frac{M_x \vec{i} + M_y \vec{j}}{T_2} - \frac{(M_z - M_0)\vec{k}}{T_1}, \qquad (3.1)$$

where $\vec{i}$, $\vec{j}$, $\vec{k}$ are the unit vectors of the Cartesian system, M_0 is the magnetization at thermal equilibrium in the presence of the static field $\vec{B}_0$, and $\vec{B}_{\text{eff}}$ is the effective magnetic field experienced by the bulk magnetization vector. T_1 and T_2 are time constants characterizing the relaxation processes undergone by a spin system perturbed from its thermal equilibrium by an RF pulse. T_1 is known

Microscopic Magnetic Resonance Imaging: A Practical Perspective
Luisa Ciobanu
Copyright © 2017 Pan Stanford Publishing Pte. Ltd.
ISBN 978-981-4774-71-0 (Paperback), 978-981-4774-42-0 (Hardback), 978-1-315-10732-5 (eBook)
www.panstanford.com

as the spin-lattice or longitudinal relaxation time and characterizes the recovery of the longitudinal magnetization. The destruction of the transverse magnetization, which happens as the spins arrive at equilibrium among themselves, is characterized by the transverse relaxation time, the time constant T_2.[a] The effect of these two relaxation processes on the image contrast will be discussed in Chapter 4.

If we neglect the relaxation processes and solve Eq. 3.1 we find the magnetization at time t after the application of the RF pulse:

$$\vec{M}(t) = M_0[\cos{(\omega t)}\vec{i} + \sin{(\omega t)}\vec{j}]. \tag{3.2}$$

Eq. 3.2 can be expressed in complex number notation:

$$M_{xy}(t) = M_x(t) + i\,M_y(t) = M_0 e^{-i\omega t}. \tag{3.3}$$

It follows that the detected MR signal has the following form:

$$S(t) = S_0 e^{-i\omega t}, \tag{3.4}$$

where S_0 is proportional to M_0.

3.2 The $\vec{k}$-space

As previously seen in Chapter 2, generating magnetic resonance images involves the application of magnetic field gradients in addition to the static and the RF fields. These magnetic field gradients will render the Larmor frequency position dependent. Let us consider the nuclear spins located at a position $\vec{r}$ in the sample, occupying a small volume of element dV. If the local spin density is $\rho(\vec{r})$, then the number of spins in a volume element dV is $\rho(\vec{r})\,dV$ and, following Eq. 3.4, the contribution to the MRI signal of the volume element dV at position $\vec{r}$ is

$$dS(\vec{G}, t) = \rho(\vec{r})\,dV e^{-i(\gamma B_0 + \gamma \vec{G}\cdot\vec{r})t}. \tag{3.5}$$

---

[a]In practice, spins will experience an additional dephasing due to external field inhomogeneities, which will lead to a faster decay of the transverse magnetization characterized by a time constant T_2^* ($T_2^* < T_2$).

After removing the carrier signal $e^{-i\gamma B_0 t}$ and integrating Eq. 3.5 we obtain

$$S(t) = \int\limits_x \int\limits_y \int\limits_z \rho(\vec{r})e^{-i\gamma \vec{G}\cdot\vec{r}t}\, dx\, dy\, dz. \qquad (3.6)$$

One can notice that Eq. 3.6 has the form of a Fourier transform. To make this explicit, we define the *reciprocal space* vector,[b] $\vec{k}$:

$$\vec{k} = (2\pi)^{-1}\gamma\vec{G}t. \qquad (3.7)$$

One can write the signal, $S(\vec{k})$, and the spin density $\rho(\vec{k})$, as

$$S(\vec{k}) = \int\limits_x \int\limits_y \int\limits_z \rho(\vec{r})e^{-2\pi i \vec{k}\cdot\vec{r}}\, dx\, dy\, dz \qquad (3.8)$$

and respectively,

$$\rho(\vec{k}) = \int\limits_x \int\limits_y \int\limits_z S(\vec{r})e^{2\pi i \vec{k}\cdot\vec{r}}\, dx\, dy\, dz. \qquad (3.9)$$

3.3 Encoding Schemes

In order to obtain an image, one has to sample the $\vec{k}$-space and then perform a Fourier transform. The $\vec{k}$-space can be traversed by moving either in time or in gradient magnitude. The direction in which the $\vec{k}$-space is sampled is determined by the direction of the gradient $\vec{G}$. There are two main ways of mapping $\vec{k}$-space: frequency and phase encoding. When the MRI signal is acquired in the presence of a gradient the signal points are obtained along a single line in $\vec{k}$-space. Let us assume the gradient is applied on the x-axis, corresponding to the k_x-axis in $\vec{k}$-space.[c] In this case, the MR detected signal is $dS(x, t) = \rho(x)\,dxe^{-i\gamma(B_0+G_x x)t}$. Because the oscillation frequency of the signal is linearly dependent on the x location the x-axis is called frequency encoded and the associated gradient is referred to as *read* or *frequency* encoding gradient. The

---

[b]Not to be confused with the unit vector of the z-axis introduced in Eq. 3.1, also denoted by $\hat{k}$.

[c]The individual points k_x in $\vec{k}$-space do not correspond to one individual pixel, x, in the MR image; each $\vec{k}$-space point contains spatial frequency information about every pixel in the final image.

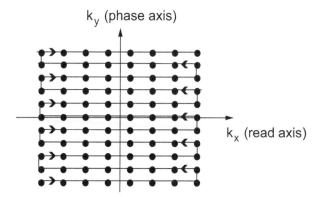

Figure 3.1 Illustration of a rectilinear sampling of $\vec{k}$-space.

intercept of the k_x-axis to the orthogonal axis k_y can be changed by applying another gradient, along the y-axis, before the next sampling begins (Fig. 3.1). As a result of this gradient, signals corresponding to spins located at different positions along y will accumulate different phase angles: $\phi(y) = -\gamma G_y y T_{pe}$, where T_{pe} is the duration of the G_y gradient. Because $\phi(y)$ depends linearly on the location y, the y-axis is *phase encoded* and the corresponding gradient is called *phase* gradient.

Assuming we average over the third dimension, z, the signal and spin density from Eqs. 3.8 and 3.9 become

$$S(k_x, k_y) = \int_x \int_y \rho(x, y) e^{-2\pi i (k_x x + k_y y)} \, dx \, dy \tag{3.10}$$

$$\rho(x, y) = \int_{k_x} \int_{k_y} S(k_x, k_y) e^{2\pi i (k_x x + k_y y)} \, dk_x \, dk_y. \tag{3.11}$$

Unless one images very thin samples (cell cultures or brain slices, for example) averaging over the entire sample in the third dimension is not desirable. This is usually avoided by selecting 2D sections (*slices*) within the sample. To selectively excite the spins in a slice, one needs a frequency-selective RF pulse and a magnetic field gradient. The gradient, called *slice-selection* gradient, is applied in a direction perpendicular to the slice plane, z in our case, and renders the spin resonance frequency position dependent. For a given gradient amplitude, G_{ss}, the range of resonance frequencies for the spins located within a slice with thickness, thk, is given by $\Delta f = \gamma G_{ss} thk$. Therefore the transmit bandwidth of the RF pulse should be Δf. In practice, Δf is held constant and the slice thickness is varied by adjusting G_{ss}. The main disadvantage of 2D imaging is the time required for slice selection. During this time, the spins are subject to transverse relaxation, which can be problematic for samples with short relaxation times. In addition, in magnetic resonance microscopy acquiring very thin slices may not be possible due to gradient strength limitations as thinner slices require stronger gradients. An alternative solution, very common in MRM, is to employ 3D imaging. In 3D imaging, a very short, "hard," pulse is applied to excite the entire sample and phase encoding is performed on both the y and z directions. The disadvantage of the 3D imaging is the increase in the acquisition time, as we will discuss later.

3.4 Image Resolution

The field of view (width) of the image and the nominal spatial resolution in the real space are dictated by the range and density of points sampled in the $\vec{k}$-space. A resolution Δx requires that the width of $\vec{k}$-space sampled along x be $W_{kx} = 1/\Delta x$. For a given field of view, FOV, the spacing between $\vec{k}$-space points is dictated by Nyquist's sampling theorem[d]: $\Delta k_x = 1/FOV$. We also recall the relation $k = \gamma Gt$ between the $\vec{k}$-space points, the gradient amplitude G and the duration t of the gradient pulses. These relations are summarized in Fig. 3.2. In practice, the true image resolution is affected by two additional factors: molecular diffusion and natural linewidth of the proton resonance (Callaghan, 1991). Minimizing the effects induced by these two factors requires large magnetic field gradients, which in turn will lead to large receiver bandwidths deteriorating the SNR (see Eq. 3.19 in the next section).

---

[d]Nyquist's sampling theorem states that a given frequency must be sampled at least twice per cycle.

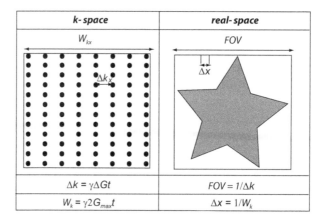

k-space	real-space
W_{kx}	FOV
$\Delta k = \gamma \Delta Gt$	$FOV = 1/\Delta k$
$W_k = \gamma 2 G_{max} t$	$\Delta x = 1/W_k$

Figure 3.2 Relations between the real and $\vec{k}$-space parameters.

3.5 Image Signal-to-Noise-Ratio

The signal-to-noise-ratio (SNR), defined as the mean signal in a region of interest divided by the standard deviation of the background noise, is a measure of the quality of the image. The SNR depends in a complex manner on the hardware and sample characteristics as well as on the acquisition parameters. In this section, we derive analytical SNR expressions which can be used to understand the interplay between these factors and to optimize MR protocols.

To calculate the SNR in the image domain, we start by evaluating the SNR in the time domain. The MR signal S from a voxel inside the sample is proportional to its volume, V_{voxel} and the B_1 field per unit current (Hoult and Richards, 1976):

$$S = \frac{1}{\sqrt{2}} \omega_0 B_1 M_0 V_{voxel}. \tag{3.12}$$

M_0 is the magnitude of the magnetization at equilibrium given by

$$M_0 = \frac{\rho \gamma^2 \hbar^2 B_0}{4 k_B T_{sample}}, \tag{3.13}$$

where ρ is the spin density (the number of spins per unit volume), T_{sample} is the temperature of the sample, $\hbar$ is the reduced Planck constant ($\hbar = 1.05457 \times 10^{-34}$ Js) and k_B is Boltzmann's constant ($k_B = 1.38 \times 10^{-23}$ JK^{-1}).

The noise level is proportional to the effective resistance of the system which is determined by the random fluctuations in the electronics and the sample (Hoult and Richards, 1976). For an effective resistance R_{eff} the variance of the noise voltage can be written as:

$$\sigma^2 = 4k_B T_{eff} BW R_{eff} \tag{3.14}$$

where T_{eff} is the temperature characterizing the noise of the system and BW is the bandwidth of the receiver.

Combining Eqs. 3.12 and 3.14 and replacing M_0 by the expression in Eq. 3.13 we obtain

$$SNR_{time} = \frac{\gamma^3 \hbar^2 B_0^2 B_1 \rho V_{voxel}}{4k_B T_{sample}} \cdot \frac{1}{\sqrt{8k_B T_{eff} BW R_{eff}}}. \tag{3.15}$$

Following Hill and Richards (1968) we define the filling factor of the coil as

$$\eta = \frac{\underset{sample}{\int} B_1^2 \, dV}{\underset{allspace}{\int} B_1^2 \, dV}. \tag{3.16}$$

The integral in the numerator represents the magnetic energy stored in the sample volume given by

$$\frac{1}{2\mu_0} \underset{sample}{\int} B_1^2 dV = \frac{B_1^2 V_{sample}}{2\mu_0}. \tag{3.17}$$

The denominator can be found from the expression for the coil inductance, L:

$$L = \frac{1}{\mu_0} \underset{allspace}{\int} B_1^2 dV \tag{3.18}$$

In Eqs. 3.17 and 3.18 μ_0 is the magnetic permeability of free space, $\mu_0 = 4\pi \times 10^{-7}$ H m^{-1}.

Substituting Eqs. 3.17 and 3.18 in Eq. 3.16 we obtain $B_1 = \sqrt{\eta \mu_0 L / V_{sample}}$, which plugging into Eq. 3.15 gives

$$SNR_{time} = \frac{\gamma \hbar^2 \omega_0 \rho V_{voxel}}{4k_B T_{sample}} \cdot \sqrt{\frac{\eta \mu_0 \omega_0 Q}{8k_B T_{eff} BW V_{sample}}}. \tag{3.19}$$

Note that in Eq. 3.19 we replaced $\omega_0 L/R_{eff}$ with Q, the quality factor of the coil, discussed in Section 2.2.3. For the special case of a solenoid and assuming uniform B_1 field, it can be demonstrated that $\eta = V_{sample}/2V_{coil}$. Therefore, for a solenoid Eq. 3.21 becomes

$$
\text{SNR}_{\text{time}}^{solenoid} = \frac{\gamma\hbar^2\omega_0\rho V_{\text{voxel}}}{4k_B T_{\text{sample}}} \cdot \sqrt{\frac{\mu_0\omega_0 Q}{16k_B T_{\text{eff}} B W V_{\text{coil}}}}. \tag{3.20}
$$

Under the assumption that the data are sampled on a Cartesian matrix, the variance of the noise in the image domain can be expressed as $\sigma_{im}^2 = \sigma^2/N$ (Haake, 1999) which implies that the $\text{SNR}_{\text{image}} = \text{SNR}_{\text{time}}\sqrt{N}$, where N is the number of $\vec{k}$-space points acquired in each scan. For a three-dimensional acquisition $N = N_x N_y N_z N_{av}$, where N_{av} represents the number of averages for the entire experiment and N_x, N_y, and N_z are the number of points acquired in the x, y, and z directions, respectively. Based on Eq. 3.19, we obtain the following expression for the $\text{SNR}_{\text{image}}$:

$$
\text{SNR}_{\text{image}} = \frac{\gamma\hbar^2\omega_0\rho V_{\text{voxel}}}{4k_B T_{\text{sample}}} \cdot \frac{\sqrt{\eta\mu_0\omega_0 Q}\sqrt{N_x N_y N_z N_{av}}}{\sqrt{8k_B T_{\text{eff}} B W V_{\text{sample}}}}. \tag{3.21}
$$

Defining the total acquisition time, T_{acq}, as the product between the number of averages, N_{av}, the number of phase encoded points, $N_y N_z$ (assuming the frequency encoding is performed along the x direction), and the repetition time, TR^e, we obtain

$$
\text{SNR}_{\text{image}} = \frac{\gamma\hbar^2\omega_0\rho V_{\text{voxel}}}{4k_B T_{\text{sample}}} \cdot \frac{\sqrt{\eta\mu_0\omega_0 Q}\sqrt{T_{acq}}\sqrt{N_x}}{\sqrt{8k_B T_{\text{eff}} B W V_{\text{sample}}}\sqrt{TR}}. \tag{3.22}
$$

3.6 Choice of Imaging Parameters

Imaging parameters are factors which can be chosen in an experiment and do not depend on the sample (spin density, relaxation times, temperature) or the MRI system (B_0, B_1). These parameters include: number of encoding points, repetition time, FOV, number of averages, etc. The SNR of the final image depends on these parameters, which must therefore be selected accordingly.

---

[e] TR is defined as the time between the application of two successive excitation pulses.

Table 3.1 Change in the image SNR and acquisition time when the voxel size is reduced by a factor of 8 (halved in each direction)

	Voxel SNR	Acquisition time	Acquisition time for constant SNR
	Voxel → Voxel/8	Voxel → Voxel/8	Voxel → Voxel/8
2D acquisition	$\times \dfrac{1}{4}$	$\times 4$	$\times 64$
3D acquisition	$\times \dfrac{\sqrt{2}}{4}$	$\times 8$	$\times 64$

Combining all non-imaging parameters into a constant C we can write Eqs. 3.21 and 3.22 in the following simplified forms:

$$\text{SNR}_{\text{image}} = C\,V_{\text{voxel}}\frac{\sqrt{N_x N_y N_z N_{av}}}{\sqrt{BW}} \qquad (3.23)$$

$$\text{SNR}_{\text{image}} = C\,V_{\text{voxel}}\frac{\sqrt{T_{\text{acq}}}\sqrt{N_x}}{\sqrt{BW}\sqrt{TR}}. \qquad (3.24)$$

It is clear from Eq. 3.21 that the SNR of an image dramatically decreases when increasing the spatial resolution. Table 3.1 shows the change in SNR per voxel and in the acquisition time when the voxel size is reduced by half in all three directions for 2D and 3D acquisitions, assuming the bandwidth and the repetition time are kept constant.

Note that for a given voxel size the SNR and the acquisition time are different for 2D and 3D acquisitions as indicated by Eqs. 3.21 and 3.22. While providing higher SNR, 3D acquisitions are more time consuming because each point on the second phase encoding direction is acquired in one repetition time. This leads to the conclusion that requiring high spatial resolution, high SNR and a large FOV is not compatible with short acquisition times. If shorter acquisitions are needed at the expense of spatial resolution and SNR, then fast encoding sequences, capturing much or all of $\vec{k}$-space in single scans, are desirable.

Chapter 4

Acquisition Strategies

In the first part of this chapter we present an overview of some of the pulse sequences used to generate MR images, discussing their advantages and disadvantages for MR microscopy. The second part is dedicated to image contrast generation, with particular focus on diffusion weighted imaging.

4.1 Pulse Sequences

There is no fundamental physics limitation which prevents the application of any MRI pulse sequence to MRM studies. However, as we will discuss in this chapter, certain approaches are less suitable than others, and most of the time MRM requires specific pulse sequence optimization.

Typically MR pulse sequences are classified according to the image contrast they produce: spin density, T_1, T_2, diffusion, perfusion, etc. In what follows, we will use an alternative classification based on the type of the signal acquired: spin echo, gradient echo, or hybrid.

4.1.1 Spin Echo (SE)

One of the most common MR imaging pulse sequences, the spin echo, is based on generating an echo signal by using multiple

Microscopic Magnetic Resonance Imaging: A Practical Perspective
Luisa Ciobanu
Copyright © 2017 Pan Stanford Publishing Pte. Ltd.
ISBN 978-981-4774-71-0 (Paperback), 978-981-4774-42-0 (Hardback), 978-1-315-10732-5 (eBook)
www.panstanford.com

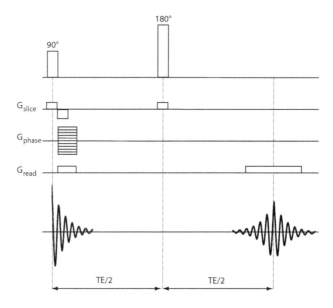

Figure 4.1 Schematic of a spin echo pulse sequence.

RF pulses. In its simplest form, the echo is generated with only two RF pulses, a 90° pulse (*excitation*) followed by an 180° pulse (*refocusing*). The 90° pulse will tip the magnetization into the transverse plane. In practice the B_0 field is never perfectly homogeneous and consequently spins at different positions will precess at slightly different frequencies. As a result, following the application of the 90° pulse, they will begin to dephase relative to each other. After a certain time interval, denoted TE/2 (Fig. 4.1), the 180° pulse is applied. The spins which have accumulated extra positive phase will now have the negative of that phase, such that the faster spins will "catch" the slower spins after another time interval TE/2 and an echo is formed. The total time between the application of the 90° pulse and the echo formation is called echo time and is typically denoted TE. The echo magnitude is a factor of $e^{-\frac{TE}{T_2}}$ smaller than the maximum amplitude of the free induction decay,[a] where T_2 is the transverse relaxation time. The popularity of spin echo sequences is justified by their robustness to B_0 inhomogeneities.

---

[a]Free induction decay (FID) is the signal immediately following the 90° pulse.

If the 90° pulse is followed by a series of 180° pulses, then a train of spin echoes, known as the Carr–Purcell–Meiboom–Gill (CPMG) echo train, will be generated (Carr, 1954; Meiboom, 1954). Typically the 180° pulses are spaced uniformly with the nth pulse applied at the time $(2n-1)\text{TE}/2$. As a result, echoes with amplitudes $E_n = e^{-n\frac{TE}{T_2}}$ will be produced at times $t_n = n\text{TE}$. When performing imaging, the signal corresponding to the different echoes can be added in order to increase the SNR. This strategy has been used in magnetic resonance microscopy and produced images with very high spatial resolution (3.5 μm isotropic) (Ciobanu, 2002). Another advantage of using CPMG-based acquisitions is that they reduce diffusion losses as a result of consecutive RF refocusing (Carr, 1954) further increasing the SNR and decreasing the resolution limit imposed by diffusion.

4.1.1.1 Fast spin echo acquisitions

While spin echoes are adequate for MRM imaging, when employed with conventional Cartesian sampling they demand long acquisition times: only one $\vec{k}$-space line is acquired in one repetition time. For example, imaging over a FOV of 2.5 mm^3 with 25 μm isotropic resolution with a repetition time, TR, of 3 s takes approximately 8 h. Fast spin echo imaging was introduced by Henning in 1986 (Henning, 1986). The method, called RARE (Rapid Imaging with Refocused Echoes), uses the CPMG approach to generate multiple echoes, each echo encoding different lines in $\vec{k}$-space (Fig. 4.2). The number of echoes generated dictates the imaging speed improvement and is called *acceleration factor*. Using the previous example, if in one TR one acquires four echoes instead of one with the proper $\vec{k}$-space encoding, the acquisition time will decrease by a factor of 4 (from 8 to 2 h). An example of an MRM RARE image is shown in Fig. 4.3.

The main limitation of RARE is image blurring along the phase encoding direction due to T_2 decay between the first and the last acquired echoes. This artifact can be reduced by keeping the duration of the echo train short relative to the T_2 decay of the species to be imaged.

Figure 4.2 $\vec{k}$-space sampling in a RARE acquisition with acceleration factor 4. First echo ·—·—·; second echo – – – ; third echo ——; fourth echo - - - - -.

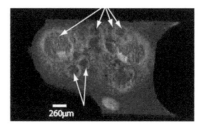

Figure 4.3 RARE image of the abdominal ganglion of *Aplysia californica* in which individual neurons can be identified (indicated by arrows). Spatial resolution 26 μm isotropic. Operating frequency 730 MHz. The acquisition parameters are listed in the Appendix.

4.1.2 Gradient Echo (GE)

To form an echo one can apply magnetic field gradients to dephase and rephase the MR signal in a controlled manner instead of using RF pulses. In this case, a negative gradient is played after the 90° RF pulse inducing a loss in spin phase coherence. If at time TE/2, at which the signal decays to zero, a positive gradient with the same magnitude is applied, the transverse magnetization will rephase and form an echo at a time TE (Fig. 4.4). The gradient echo will refocus only the dephasing introduced by the applied magnetic field gradient and, unlike the spin echo, will not fully refocus the spins dephased due to B_0 inhomogeneities. In consequence, the echo formed will be T_2^* instead of T_2 weighted. At high magnetic

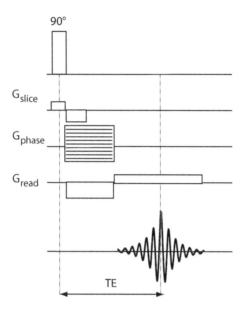

Figure 4.4 Schematic of a gradient echo pulse sequence.

fields, magnetic susceptibility differences, both macroscopic and microscopic, cause a significant reduction in T_2^*. While this leads to an increased contrast caused by iron deposits or the presence of blood, it can also cause image artifacts and decrease the image SNR. The latter consequence makes the gradient echo sequences less suitable for high-field MR microscopy than the spin echo methods discussed earlier.

Gradient echo sequences are often used in combination with small excitation flip angles. A flip angle lower than 90° decreases the amount of magnetization tipped into the transverse plane requiring a shorter time for the recovery of the longitudinal magnetization allowing, therefore, shorter repetition times for fast imaging. The basic GE sequences employ repetition times shorter than T_1 but longer than T_2.

4.1.2.1 Fast gradient echo acquisitions

Fast GE methods constitute a particular class of GE imaging sequences with repetition times shorter than the T_2 relaxation time.

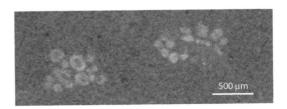

Figure 4.5 Selected slice from a 3D FLASH image of the buccal ganglia of *Aplysia californica*. The hyperintense regions represent individual neurons. Spatial resolution 25 µm isotropic. Operating frequency 730 MHz. The acquisition parameters are listed in the Appendix.

In fast GE acquisitions, the transverse magnetization M_{xy} is non-zero at the application of a new RF-pulse and therefore contributes to the final steady-state signal intensity. As a result, these sequences provide higher SNR than the basic GE sequences but the contrast depends in a complex fashion on both T_1 and T_2. The T_1 contrast can be enhanced while maintaining very short repetitions times by spoiling the transverse component of the magnetization via RF (SPGR, FLASH, T1-FFE) or gradient (GRASS, FISP, FAST) spoiling (Hargreavis, 1986).

The most common gradient echo fast imaging method is known as Fast Low Angle SHot (FLASH) imaging and was initially proposed by Haase et al. (1986). Despite a reduction in signal intensity, and therefore SNR, due to the low flip angle and short T_2^*, FLASH acquisitions are sometimes used in MRM in order to obtain T_1-weighted images, as we demonstrate in Fig. 4.5.

4.1.3 Hybrid Pulse Sequences

Hybrid pulse sequences use both spin echoes and gradient echoes, the most popular one being the echo planar imaging (EPI) technique, introduced by Mansfield and co-workers (1977). EPI is a fast acquisition method in which one entire $\vec{k}$-space raster is acquired in one single TR. Following the 90°/180° pulses, multiple echoes are acquired using rephasing gradients produced by rapidly reversing the frequency encoding gradient. The most commonly used $\vec{k}$-space acquisition scheme is shown in Fig. 4.6 (*blipped* EPI). EPI sequences can also be used in full gradient echo mode, where the 90°/180° pair

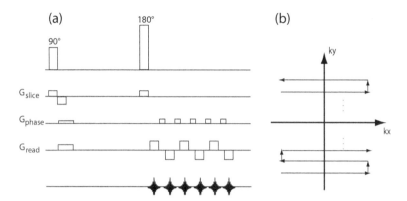

Figure 4.6 (a) Blipped EPI pulse sequence diagram (b) 2D EPI $\vec{k}$-space raster

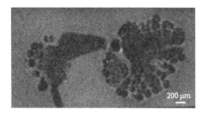

Figure 4.7 Selected slice from 3D SE-EPI image of the pedal ganglia of *Aplysia californica*. Spatial resolution: 20 μm isotropic. Operating frequency 730 MHz. The acquisition parameters are listed in the Appendix.

is replaced by one single excitation pulse (with a flip angle smaller than 90°). One particularity of the EPI sequence is the need for rapid gradient switching. For a successful EPI experiment, one therefore needs screened gradient coils which prevent induced eddy currents. While technically possible (Fig. 4.7), the use of EPI in MR microscopy is very limited, restricted by the low image SNR caused by the large acquisition bandwidths required and by the high sensitivity to B_0 inhomogeneities.

4.1.4 Accelerated Acquisitions

Even with fast spin echoes, the acquisition times required for very-high-resolution imaging can be long, precluding the analysis of live biological samples. One way to further reduce the data acquisition time is to undersample the $\vec{k}$-space, a strategy proposed by several methods including partial Fourier data acquisition, parallel imaging and compressed sensing.

Partial Fourier acquisitions rely on the fact that the Fourier transform of a real function has complex conjugate symmetry in $\vec{k}$-space. In theory this means that it is sufficient to acquire one half of the $\vec{k}$-space points reducing therefore the acquisition time by a factor of two. The main drawbacks of partial Fourier imaging are the loss in SNR due to the reduced number of acquisition points and the enhanced sensitivity to artifacts due to the duplication of the $\vec{k}$-space data.

Parallel imaging exploits redundancy in $\vec{k}$-space reconstructing the image from data acquired simultaneously with an array of RF coils (Griswold, 2002; Pruessmann, 1999). Phased array microcoils have been developed thanks to advances in wire bonding technology (Gbel, 2015; Gruschke, 2012). However, as mentioned in Chapter 2, the small sample size renders the construction of such micro-arrays difficult and limits the applicability of parallel imaging to high-resolution MR microscopy.

Compressed sensing (CS) is a signal processing technique introduced by Donoho in 2006 (Donoho, 2006). CS produces images from significantly fewer data points than what is required by the Nyquist criterion using a non-linear reconstruction which enforces both sparsity of the image representation and consistency with the acquired data. Taking into consideration the MR hardware constrains several ways of generating undersampling patterns have been proposed. The most commonly used undersampling schemes, either Cartesian or non-Cartesian, consist of variable-density random trajectories (Lustig, 2007) based on a probability density function. Nguyen et al. (2015) introduced a new method to generate undersampling patterns based on the diffusion limited aggregation (DLA) random growth model and applied it to the undersampling of RARE and FLASH pulse sequences. In both cases, the DLA acquisition scheme reduced the experimental time by 50% and was used to perform magnetic resonance microscopy on *Aplysia* ganglia (Fig. 4.8).

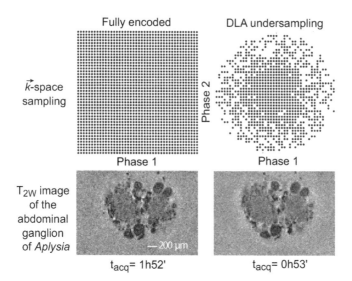

Figure 4.8 Fully encoded (left) and 50% DLA accelerated (right) *k*-space sampling and the corresponding images of the abdominal ganglion of *Aplysia californica*. Image resolution 25 μm isotropic. Operating frequency 730 MHz. Reprinted from Nguyen et al. (2015) with permission from Elsevier.

Besides anatomical imaging, compressed sensing strategies can also be applied to quantitative MRI studies including T_1, T_2 and T_2^* relaxation times (Doneva, 2010) and diffusion (McClymont, 2016). Such studies are becoming widespread in clinical and preclinical research fields. In MR microscopy, CS applications are still somewhat isolated; however, we expect these to grow in the future.

4.2 Contrast Mechanisms

4.2.1 Basic Contrasts

There are three main types of contrast in MR imaging: spin density, T_1 and T_2/T_2^*. The measured signal intensity is a function of all these three characteristics of the sample. By appropriately choosing the acquisition parameters one can bring forward a particular contrast over another (*weighting*). In order to examine the image contrast generated by a specific pulse sequence one has to derive the corresponding expression for the signal. Let us take, for example, a simple spin echo experiment with a repetition time TR (the time between two successive pairs of 90°/180° pulses). Solving the Bloch equation, Eq. 3.1, in the rotating frame ($\vec{B}_{\text{eff}} = 0$), we obtain the following solutions for the transverse and longitudinal magnetization components:

$$M_{xy}(t) = M_{xy}(0)e^{-t/T_2} \tag{4.1}$$

$$M_z(t) = M_0(1 - e^{-t/T_1}) + M_z(0)e^{-t/T_1}, \tag{4.2}$$

where $M_{xy}(0)$ and $M_z(0)$ are the magnetizations in the transverse plane and along the z axis immediately after the application of the RF pulse and M_0 is the magnetization at thermal equilibrium. At time TE/2 after the application of the 90° pulse, right before the application of the 180° pulse, the z magnetization is

$$M_z(\text{TE}/2) = M_0(1 - e^{-\text{TE}/2T_1}). \tag{4.3}$$

After the application of the 180° pulse the z component of the magnetization changes sign and becomes

$$M_z(\text{TR}) = M_0(1 - 2e^{-(\text{TR}-\text{TE}/2)/T_1} + e^{-\text{TR}/T_1}), \tag{4.4}$$

after time TR, at the application of the second 90° pulse when it gets tipped into the transverse plane and constitutes the initial value of the transverse magnetization which, at the echo, has the following magnitude:

$$M_{xy} = M_0 \left(1 - 2e^{-(\text{TR}-\text{TE}/2)/T_1} + e^{-\text{TR}/T_1}\right) e^{-\text{TE}/T_2}. \tag{4.5}$$

Typically TE $\ll$ TR and Eq. 4.5 reduces to

$$M_{xy} = M_0 \left(1 - e^{-\text{TR}/T_1}\right) e^{-\text{TE}/T_2}. \tag{4.6}$$

Correspondingly, the amplitude of the echo can be written as

$$S_{\text{SE}} \propto M_0 \left(1 - e^{-\text{TR}/T_1}\right) e^{-\text{TE}/T_2} \propto \rho \left(1 - e^{-\text{TR}/T_1}\right) e^{-\text{TE}/T_2}, \tag{4.7}$$

where ρ is the spin density. A similar expression can be written for a gradient echo pulse sequence, by replacing T_2 with T_2^* (assuming a 90° flip angle).

$$S_{\text{GE}} \propto \rho \left(1 - e^{-\text{TR}/T_1}\right) e^{-\text{TE}/T_2^*} \tag{4.8}$$

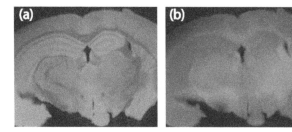

Figure 4.9 Images of a chemically fixed slice of a mouse brain showing (a) T_2 and (b) T_1 contrast with the white matter appearing hypo- and hyperintense, respectively. In plane resolution 30 μm, operating frequency 730 MHz. The acquisition parameters are listed in the Appendix.

For a gradient echo sequence with an arbitrary flip angle α, it can be shown that the echo amplitude is

$$S_{\mathrm{GE}}^{\alpha} \propto \frac{\rho(1 - e^{-\mathrm{TR}/T_1})}{1 - \cos\alpha\, e^{-\mathrm{TR}/T_1}} \sin\alpha\, e^{-\mathrm{TE}/T_2^*}. \tag{4.9}$$

It is clear from Eqs. 4.7–4.9 that by using sequences with very short echo times and very long repetition times one obtains spin density-weighted images. A spin echo sequence can produce T_2-weighted images by using long echo times and long repetition times. Similarly, a gradient echo sequence with long echo times yields T_2^*-weighted images. The T_1-contrast is dependent on the repetition time in spin echo sequences, and on the flip angle and repetition time in gradient echo sequences.

One of the advantages of MRI compared to other imaging techniques (such as computed tomography, for example) is the excellent soft tissue contrast it offers. White and gray matter brain tissue have different relaxation times based on which they can be differentiated from each other (Fig. 4.9). Approximate values for white and gray matter relaxation times at 17 T are: $T_1^{\mathrm{gray}} = 2$ s, $T_1^{\mathrm{white}} = 1.8$ s, $T_2^{\mathrm{gray}} = 25$ ms, $T_2^{\mathrm{white}} = 20$ ms.

4.2.1.1 MR contrast agents

In some situations exogenous compounds are used to enhance a particular contrast. These compounds are paramagnetic and superparamagnetic molecules which locally modify the magnetic environment of the spins and change their relaxation times.

Supermagnetic contrast agents usually consist of iron oxides (magnetite, maghemite) and their predominant effect is to shorten T_2/T_2^* reducing therefore the signal intensity in T_2/T_2^*-weighted images (negative contrast agents). Paramagnetic contrast agents (complexes of gadolinium or manganese, for example) are used to shorten the T_1 relaxation time and produce hyperintensities in T_1-weighted images (positive contrast agents). The extent to which a contrast agent, either positive or negative, changes the respective relaxation time of the tissue (T_1 or T_2) is determined by its relaxivity (r_1 or r_2) and its concentration, [CA], according to:

$$\frac{1}{T_i} = \frac{1}{T_i^0} + r_i[\text{CA}], \quad i = 1, 2, \tag{4.10}$$

where T_i and T_i^0 are the relaxation times in the presence and the absence of the contrast agent, respectively.

The relaxivity depends on the physicochemical properties of the contrast agent (composition, size, surface properties) but also on external factors such as the external magnetic field and the temperature. The relaxivity of MR contrast agents is usually measured in water (at specified field strength and temperature); however, it should be noted that the effective in vivo relaxivities may differ from the in vitro values. Assuming the relaxivity of the contrast agent is known, one can determine its local concentration from Eq. 4.10 and by measuring the relaxation times.

4.2.2 Diffusion-Weighted Imaging

Besides the three basic types of contrast discussed in the previous section, MR images can be sensitized to other physical processes such as molecular diffusion, perfusion, oxygenation level, etc. Among these, diffusion imaging is often used in magnetic resonance microscopy. The evolution of magnetization in a magnetic field in the presence of molecular diffusion is described by the Bloch–Torrey equation, which is obtained by including an additional *diffusion* term to the Bloch equation:

$$\frac{d\vec{M}}{dt} = \gamma \vec{M} \times \vec{B} + D\nabla^2 \vec{M}, \tag{4.11}$$

where D is the diffusion coefficient. Assuming a spatially invariant diffusion coefficient, the solution of Eq. 4.11 can be expressed as

$$M_{xy}(t) = A(t)e^{-i\gamma \int_0^t \vec{G}(t')\cdot\vec{r}\,dt'}. \tag{4.12}$$

Substituting Eq. 4.12 in Eq. 4.11, we find the following solution for $A(t)$:

$$\ln A(t) = -D\gamma^2 \int_0^t dt' \left[\left(\int_0^{t'} \vec{G}(t'')\,dt'' \right) \cdot \left(\int_0^{t'} \vec{G}(t'')\,dt'' \right) \right] + \ln A(0) \tag{4.13}$$

The factor $b = \gamma^2 \int_0^t dt' \left[\left(\int_0^{t'} \vec{G}(t'')\,dt'' \right) \cdot \left(\int_0^{t'} \vec{G}(t'')\,dt'' \right) \right]$, is referred to as *b-value* or *diffusion weighting*. For a gradient waveform satisfying the condition $\int_0^t \vec{G}(t') \cdot dt' = 0$, we obtain the following expressions for the magnetization and the MR signal:

$$M_{xy} = A(0)e^{-bD} \tag{4.14}$$

and respectively,

$$S = S_0 e^{-bD}. \tag{4.15}$$

In the basic diffusion-weighted MRI sequence, known as pulsed gradient spin echo (PGSE), the diffusion weighting is accomplished by inserting a gradient pulse, G_{diff}, on either side of the 180° RF pulse (Fig. 4.10). For the diffusion gradients illustrated in the pulse sequence diagram of Fig. 4.10 the b-value can be shown to be $b = \gamma^2 G_{diff}^2 \delta^2 (\Delta - \delta/3)$ and the MR signal can be written as

$$S = S_0 e^{-\gamma^2 G_{diff}^2 \delta^2 D(\Delta - \delta/3)}. \tag{4.16}$$

Equation 4.16 was obtained considering the rectangular diffusion gradient pulses displayed in Fig. 4.10. In practice, corrections due to finite gradient pulse raise times and imaging gradients have to be applied. We also recall that when solving Eq. 4.11, we made the assumption that diffusion is isotropic and unrestricted. However, in biological tissues diffusion is most of the time anisotropic and restricted. The MR signal attenuation can still be modeled using Eq. 4.15, in which the diffusion coefficient D is replaced by an apparent diffusion coefficient, ADC:

$$S = S_0 e^{-b ADC}. \tag{4.17}$$

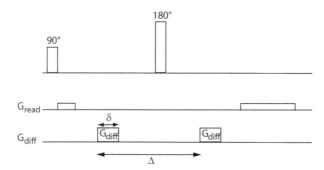

Figure 4.10 Simplified schematic of a standard PGSE sequence. The diffusion encoding gradients can be applied in any direction by using a combination of x, y, and z gradients.

The measured ADC depends on the tissue microstucture (Le Bihan, 2007) and on the experimental parameters. In particular, ADC significantly changes with the diffusion time, defined as $\Delta - \delta/3$, which dictates the duration over which the diffusion is encoded. At very short times ADC reflects the intrinsic local viscosity. At long diffusion times ADC is a measure of the integrated effects of all the restrictions encountered by the water molecules and depends on the tissue properties such as membrane permeability and cell density. From these dependencies one can obtain specific structural information such as the size of different neuronal components (neurons, axons) by carefully choosing the diffusion encoding time. According to the Einstein equation (Einstein, 1905), to probe spherical cells with diameters of 7 μm one needs diffusion encoding times on the order of 10 ms considering the intrinsic diffusion coefficient of free water (2.5×10^{-5} cm^2/s). In this regime, for the short gradient pulse approximation (δ is so short that the spins are assumed not to diffuse during the pulse duration), porous media studies showed that the ADC is given by (Mitra, 1993)

$$ADC_{\text{short}} = D_0 \left(1 - \frac{4\sqrt{D_0}}{3d\sqrt{\pi}} \sqrt{\Delta} \frac{S}{V} \right), \qquad (4.18)$$

where D_0 is the free diffusivity, S/V is the surface-to-volume ratio and d is the number of dimensions.

When taking into account the finite pulse duration, Eq. 4.18 becomes (Schiavi, 2016):

$$ADC_{\text{short}} = D_0 \left(1 - \frac{4\sqrt{D_0}}{3d\sqrt{\pi}} C_{\delta,\Delta} \frac{S}{V} \right),$$ (4.19)

where

$$C_{\delta,\Delta} = \frac{4}{35} \frac{(\Delta+\delta)^{7/2} + (\Delta-\delta)^{7/2} - 2\left(\delta^{7/2} + \Delta^{7/2}\right)}{\delta^2 \left(\Delta - \delta/3\right)}$$ (4.20)

$$= \sqrt{\Delta} \left(1 + \frac{1}{3}\frac{\delta}{\Delta} - \frac{8}{35}\left(\frac{\delta}{\Delta}\right)^{3/2} + \cdots \right).$$ (4.21)

When $\delta \ll \Delta$, $C_{\delta,\Delta}$ becomes $\sqrt{\Delta}$ and Eq. 4.19 reduces to Eq. 4.18.

For spheres ($d = 3$) or disks ($d = 2$), the surface-to-volume ratio S/V is d/R where R is the radius. In practice, one can estimate R by fitting the experimental ADC as a linear function of $\sqrt{\Delta}$ or $C_{\delta,\Delta}$.

ADC also depends on the direction on which the diffusion encoding gradients are applied. Figure 4.11 shows ADC maps in which the corpus callosum (cc) and the internal capsule (ic) are hyperintense

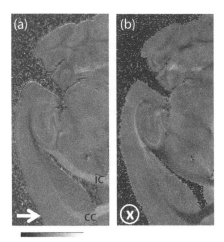

Figure 4.11 ADC maps of a chemically fixed mouse brain obtained with diffusion encoding gradients oriented (a) from left to right and (b) perpendicular to the image plane. The gray scale bar indicates an ADC range from 0.5×10^{-3} to 1.5×10^{-3} mm^2/s. Spatial resolution: 50 μm × 50 μm × 130 μm. Operating frequency 730 MHz. The acquisition parameters are listed in the Appendix.

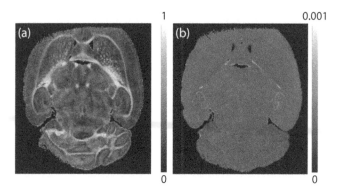

Figure 4.12 (a) FA and (b) MD maps of a chemically fixed mouse brain slice. Spatial resolution: 50 μm × 50 μm × 130 μm. Operating frequency 730 MHz. The acquisition parameters are listed in the Appendix.

or hypointense depending on the direction of the diffusion encoding gradients. It is clear that in the presence of anisotropy, diffusion cannot be characterized by a single scalar coefficient; instead it requires a tensor and therefore measurements have to be performed along several gradient directions (minimum six). From the diffusion tensor components, one can calculate different metrics providing information about tissue microarchitecture, the most common ones being mean diffusivity (MD) and fractional anisotropy (FA) (Le Bihan, 2001). Figure 4.12 shows FA and MD maps of a fixed mouse brain. The white matter structures can be clearly visualized on the FA map.

For high b-values, the decay of the diffusion-weighted signal deviates from the mono-exponential model described by Eq. 4.17. Empirically this decay was shown to be better fitted by a weighted sum of two exponential functions:

$$S_{\text{biexp}} = S_0(w_1 e^{-bD_1} + w_2 e^{-bD_2}). \tag{4.22}$$

Equation 4.22 suggests the presence of two compartments, characterized by two different diffusion coefficients, D_1 and D_2, and for which the weights, w_1 and w_2, represent the relative volume fractions. The initial interpretation of this model was that the two compartments correspond to the intra and extracellular spaces. This hypothesis was, however, not confirmed experimentally and there is still a debate regarding the physical origins of the two pools. Despite this, the biexponential description of tissue water diffusion continues to be used due to its robustness and simplicity. Other models have nonetheless been proposed including the kurtosis model (Jensen, 2005), the statistical model (Yablonskiy, 2003) and the stretched exponential model (Bennett, 2003). Among these, the most popular one is the kurtosis model in which the decay of the signal is fitted with a Taylor expansion truncated at the second term:

$$S_{\text{biexp}} = S_0 e^{-bD + \frac{K}{6}(bD)^2}. \tag{4.23}$$

In Eq. 4.23, K, the kurtosis, quantifies the deviation from the Gaussian behavior (for Gaussian diffusion $K = 0$).

4.2.2.1 Diffusion acquisitions for MR microscopy

Diffusion measurements are time consuming as they typically require multiple b-values and multiple diffusion encoding directions. For this reason, the acquisition strategy most commonly used in clinical and preclinical imaging is the diffusion-weighted EPI. However, as discussed earlier, EPI acquisitions are not favorable to magnetic resonance microscopy studies due low SNR and their inherent sensitivity to B_0 inhomogeneities. Incorporating diffusion-weighted modules in fast GE acquisitions such as FISP (fast imaging with steady-state free precession) are good alternatives to EPI. Such approaches are not suitable for diffusion quantification because the resulting signal is a complicated function of diffusion, T_1 and T_2 weightings (Deoni, 2004; Le Bihan, 1988). For straightforward diffusion quantification one can use diffusion preparation modules followed by fast acquisitions. In the schematic presented in Fig. 4.13, the diffusion weighting is imparted to the

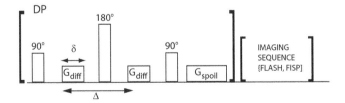

Figure 4.13 Schematic of a diffusion preparation module.

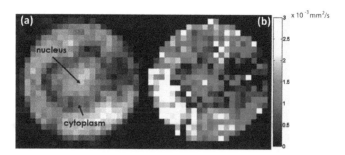

Figure 4.14 (a) Axial DP-FISP image of an *Aplysia* neuron for $b = 0$ s/mm^2 and (b) the corresponding ADC parametric map. Spatial resolution 25 μm isotropic. Operating frequency 730 MHz. The cytoplasm and the nucleus are distinguishable in both images. The acquisition parameters are listed in the Appendix. Images courtesy of Dr. Ileana Jelescu.

longitudinal magnetization in the preparation module (DP) and the image encoding is performed using fast gradient echo-based acquisitions. The main drawback of this preparation scheme is the longitudinal magnetization recovery between the end of the preparation module and the beginning of the acquisition which leads to T_1 contamination of the diffusion-weighted signal. In order to minimize this contamination, one has to encode the central $\vec{k}$-space points immediately after the diffusion preparation by performing centrically ordered encoding.

Jelescu et al. (2014) report the implementation of a 3D diffusion prepared (DP)-FISP pulse sequence with suitable timings (9 min per b-value) and resolutions (25 μm isotropic) for MR microscopy of unfixed biological tissue. Using this sequence the authors report ADC maps of single isolated neurons in which the nucleus and the cytoplasm can be clearly differentiated (see Fig. 4.14). It is worth noting that in order to avoid T_1 contamination and to ensure a correct assessment of the effective b-value, this sequence was used for low b-values (<600 s/mm^2).

Chapter 5

Image Artifacts

Artifacts are aberrant features of an image which are not present in the original imaged object. Magnetic resonance microscopy suffers from the same artifacts as magnetic resonance imaging. However, MRM artifacts are accentuated due to the strong magnetic fields typically used, the increased hardware requirements and the high spatial resolutions intended. Image artifacts can have multiple causes, including hardware imperfections, pulse sequence limitations, or improper data acquisition. In this chapter, we focus on some of the most common artifacts and, where possible, present strategies for their avoidance or minimization.

5.1 Magnetic Susceptibility Artifacts

Magnetic susceptibility artifacts are caused by large magnetic susceptibility differences at interfaces between different types of materials. Depending on their precise origin as well as on the acquisition parameters they can manifest themselves as image distortions and/or variations in signal intensity. In magnetic resonance microscopy the main causes of such artifacts are air bubbles introduced during sample preparation and the close proximity of

Microscopic Magnetic Resonance Imaging: A Practical Perspective
Luisa Ciobanu
Copyright © 2017 Pan Stanford Publishing Pte. Ltd.
ISBN 978-981-4774-71-0 (Paperback), 978-981-4774-42-0 (Hardback), 978-1-315-10732-5 (eBook)
www.panstanford.com

the electronic components (RF coil) to the object being imaged. As mentioned in Chapter 2, susceptibility artifacts caused by the proximity of the wires of the coil to the sample can be mitigated by inserting the RF coil in Fluorinert or other susceptibility matching materials. In the remainder of this chapter, we will discuss magnetic susceptibility artifacts induced by the presence within the sample of materials with very different magnetic susceptibilities such as air bubbles.

The mean magnetic field within a sample with magnetic susceptibility[a] χ_m placed in an external field B_0 is given by $B_0' = (1 + \chi_m)B_0$. A foreign object with a different magnetic susceptibility located in the sample will experience a different mean magnetic field from that of the sample itself. This difference will generate a local magnetic field gradient at its boundaries. According to Eq. 2.12, variations in the local magnetic field will induce changes in the precession frequency. As a result, the frequency position encoding will be inexact and the image will no longer accurately reflect the spatial distribution of the spins. The magnitude of this position encoding error scales as $\Delta\chi_m B_0/G$, where $\Delta\chi_m$ is the difference between the magnetic susceptibilities of the foreign object and the sample, and G is the magnitude of the imaging gradient. In the final image, the position misregistration will manifest as image shift and deformation, often with hyperintense rims around the foreign object (Fig. 5.1).

The increased magnetic field inhomogeneity not only generates the incorrect position mapping but also leads to intravoxel dephasing. This in turn reduces locally the relaxation time T_2^* causing a signal loss in gradient-echo images (spin-echo images will be impacted to a much smaller extent). The net effect will be an amplification of the size of the foreign object in the final image for T_2^*-weighted images (gradient echo acquisitions). The magnitude of this effect increases with the echo time. Figure 5.2 shows gradient echo images of a glass capillary filled with water in which an air bubble was introduced intentionally. It can be seen that the longer echo time leads to a larger signal void created by the presence of the

---

[a] χ_m is expressed in ppm. The magnetic susceptibility of tissue is close to that of water $\chi_m^{\text{water}} = -9.04$ ppm. By contrast the magnetic susceptibility of air is $\chi_m^{\text{air}} = 0.4$ ppm.

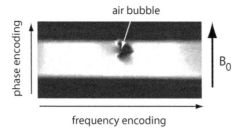

Figure 5.1 MR image of a capillary filled with water and containing an air bubble. The image was acquired using a spin echo sequence with an in plane resolution of $50 \times 50 \ \mu m^2$. Operating frequency 730 MHz. The acquisition parameters are listed in the Appendix.

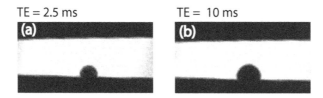

Figure 5.2 MR image of a capillary filled with water and containing an air bubble. The images were acquired using a GE sequence with an in plane resolution of $50 \times 50 \ \mu m^2$. Operating frequency 730 MHz. Compared to (a), the image in (b) uses a longer echo time which causes the bubble to appear bigger. The acquisition parameters are listed in the Appendix.

bubble. Very often, in gradient echo images susceptibility induced signal loss outweighs the image distortions described previously.

From the discussion above it is clear that intravoxel dephasing can be reduced by shortening the echo time or by using spin echo instead of gradient echo pulse sequences. The simplest way to avoid susceptibility distortions is to use large imaging gradients. Another way to avoid susceptibility artifacts is to employ pulse sequences less sensitive to large field inhomogeneity conditions. Such sequences have been proposed recently but have not been yet applied to microscopy studies (Zhang, 2016).

In spite of the problems mentioned here, susceptibility differences can sometimes be useful, as they can generate additional image contrast. Gradient echo sequences are often used to detect hemorrhages, calcifications (Wu, 2009), and iron deposits in different pathologies (Schipper, 2012). Moreover, functional MRI is based on detecting changes in magnetic susceptibility of blood during specific tasks (Ogawa, 1990). In magnetic resonance microscopy, magnetic susceptibility is used as a contrast mechanism to detect single cells loaded with iron particles at low concentrations (Dodd, 1999).

5.2 Chemical Shift Artifacts

Chemical shift artifacts are due to differences in the resonance frequencies of spins located in different chemical environments. For example, at 17.2 T (730 MHz) the resonance frequencies of water and silicone oil are separated by approximately 3000 Hz. If this difference is larger than the bandwidth associated to one pixel ($\Delta\omega = \gamma G \Delta x$, Δx representing the voxel size and G the magnitude of the imaging gradient) the correspondence between resonance frequency and position is no longer valid and will result in a displaced image (Fig. 5.3). The exact spatial shift can be calculated as $\Delta x_{ch} = \delta B_0 / G$, where the chemical shift, δ, is expressed in ppm.

Chemical shift artifacts can arise in biological samples at boundaries between fat and other tissue and they will manifest along both the frequency encoding and the slice selection directions. One way to reduce these artifacts is to increase the acquisition

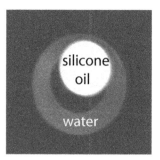

Figure 5.3 MR image of a silicone oil capillary inside a water tube. The eccentric position of the silicone oil image is caused by a chemical shift artifact. Resonance frequency 730 MHz, in plane resolution 40 μm × 40 μm. The acquisition parameters are listed in the Appendix.

bandwidth at the expense of decreasing the SNR. In practice, separated water and fat images can be obtained by selectively saturating, or exciting, one single species.

The cleanest way to image samples containing multiple chemical species is by increasing the dimensionality of the acquisition. The extra dimension is used to store the chemical shift information, in the absence of the position encoding gradients, strategy known as *chemical shift imaging (CSI)*. In this case, the spatial location is encoded using only phase encoding gradients leading to an increase in the acquisition time. For an $N \times N \times N$ matrix size image the acquisition time is increased by a factor of N.

5.3 Motion Artifacts

At very high resolutions, the smallest movement of the object being imaged will lead to image artifacts such as blurring or ghosting. There are several different types of motion which can occur during an MR scan. For in vivo preclinical or clinical imaging, the origin of the motion is most of the time physiologic (blood flow, respiration, cardiac movement). In MR microscopy motion artifacts are likely due to sample vibration caused by acoustic pulses on current gradient switching. Flow induced motion artifacts are often seen in images acquired under perfusion conditions.

The physical mechanism underlying the appearance of motion artifacts is the incorrect phase accumulation during the periods when the imaging gradients are turned on. Compared to static spins, spins moving in the direction of the encoding gradient will acquire an additional phase. When the motion takes place along the frequency encoding direction, this phase shift results in blurring and reduced signal intensity. Motion along the phase encoding direction will produce ghosting artifacts.

The best way to avoid motion artifacts is to make sure that the sample is tightly fixed inside the RF coil and the gradient assembly. When this is not possible, one should seek to minimize these artifacts as follows. Given that the sampling rate is much slower on the phase encoding than on the frequency encoding direction, motion artifacts will be a lot more accentuated when the movement

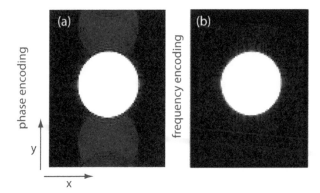

Figure 5.4 Images of a water capillary which vibrates along the *y*-direction. (a) Phase encoding is performed along the *y*-direction; the motion artifacts (ghosting) are visible. (b) Frequency encoding is performed along the *y*-direction reducing the artifacts. In plane resolution of 50 × 50 μm². Operating frequency 730 MHz. The acquisition parameters are listed in the Appendix.

takes place on the phase encoding direction. One way to reduce motion artifacts is therefore to set the frequency encoding direction in the direction of the motion as illustrated in Fig. 5.4. Signal averaging will also remove ghosting artifacts but at the expense of considerable blurring and loss of spatial resolution.

Flow induced motion artifacts can be eliminated by using flow-compensated gradient wave-shapes. This method minimizes the phase shift accumulated by the moving spins. Mathematically this is equivalent to minimizing both the 0th gradient moment ($\int_0^{\text{TE}} G(t)\, dt$) and the first gradient moment ($\int_0^{\text{TE}} vtG(t)\, dt$) at the echo time.

5.4 Aliasing Artifacts

Aliasing or wrap-around artifacts happen when the size of the imaged object is larger than the defined field-of-view (FOV) (Fig. 5.5). While aliasing artifacts occur on both phase end frequency encoding directions they are much easier to deal with on the frequency encoding direction either by applying a low-pass frequency filter or by oversampling (with no increase in the acquisition time). On

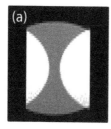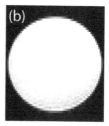

Figure 5.5 Aliasing artifact. (a) Undersampling on the phase encoding direction by a factor of 2. (b) Correct image. In plane resolution of 50 × 50 μm². Operating frequency 730 MHz. The acquisition parameters are listed in the Appendix.

the phase encoding direction, filtering is not possible as the phase increment is cyclic. Oversampling the number of data points in the phase encoding direction will remove the aliasing artifacts but will increase the acquisition time accordingly: a 50% oversampling will increase the acquisition time by 50%. Alternatively, saturation pulses can be used to eliminate the signal from regions outside the desired FOV.

5.5 RF Coil Calibration Artifacts

RF coil calibration artifacts[b] are due to the miscalibration of the RF power and manifest themselves as signal intensity variations across the image. In general, pulse sequences employing fewer RF pulses are more tolerant to pulse miscalibration. Figure 5.6a shows the effect of overestimating the necessary RF power for an image acquired using a spin echo pulse sequence. In Fig. 5.6b, the homogeneity of the image is restored by using the correct pulse power.

As discussed in Chapter 2, surface coils have inherent inhomogeneous B_1 profiles. This leads to significant variations in the RF amplitude, and therefore to unavoidable signal intensity variations across the image. One way to overcome this problem is to use pulses

---

[b]Not to be confused with RF inhomogeneity artifacts, which are caused by problems in RF construction or by dielectric effects.

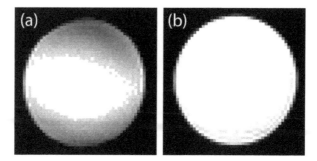

Figure 5.6 SE MR images of a capillary filled with water. The images were acquired using a solenoidal RF coil: (a) with miscalibrated RF pulses (too much power) and (b) with correct calibration. Operating frequency 730 MHz. The acquisition parameters are listed in the Appendix.

which can excite the magnetization uniformly even in the presence of large B_1 inhomogeneities (*adiabatic pulses*) (Tannus, 1997).

5.6 Clipping Artifacts

Clipping artifacts, also known as ADC (Analog-to-Digital Converter) overflow artifacts, occur when the received signal is larger than the maximum ADC value. This clipping leads to loss in low frequency data, occurring in the center of $\vec{k}$ -space where the signal reaches its maximum value. After reconstruction the image appears as a bright halo (Fig. 5.7). Clipping artifacts are easily removed by lowering the receiver gain. Most modern MR spectrometers allow for the receiver gain calibration at the beginning of the experiment, avoiding this type of artifacts.

5.7 Gibbs Ringing Artifacts

Gibbs ringing artifacts occur at sharp boundaries and large variations in image signal intensity and appear as alternating parallel dark and bright bands (Fig. 5.8). Ringing artifacts are caused by insufficient sampling at high frequencies and can appear on both frequency and phase encoding directions. The distance between

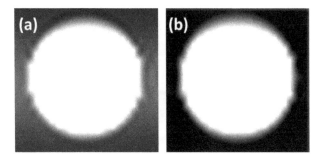

Figure 5.7 Images of a capillary filled with water (a) presenting an ADC artifact, caused by having the receiver gain set too high (b) without the artifact, after the receiver gain was correctly adjusted. Operating frequency 730 MHz. The acquisition parameters are listed in the Appendix.

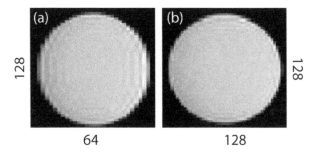

Figure 5.8 Gibbs ringing artifact. (a) Water phantom image obtained with a matrix size of 128 × 64 in which the Gibbs ringing pattern is visible. (b) The image obtained with a larger matrix size (128 × 128) shows a reduction in the artifact. Operating frequency 730 MHz. The acquisition parameters are listed in the Appendix.

alternating rings decreases with increasing the number of points and therefore these artifacts are less visible on high-resolution images where large matrices are employed.

Gibbs ringing artifacts can never be completely removed as the data sampling is always finite. One way to reduce them is by increasing the matrix size. Another way to alleviate ringing artifacts is to filter the $\vec{k}$-space data with a smoothly decreasing window prior to processing. This approach has, however, the drawback of decreasing the spatial resolution.

Figure 5.9 Image of a capillary filled with water showing a zipper artifact. The artifact was produced by an electrical stimulator. Operating frequency 730 MHz. The acquisition parameters are listed in the Appendix.

5.8 Zipper Artifacts

Zipper artifacts appear as bright lines along the phase encoding axis (Fig. 5.9). They are caused by unwanted radiofrequency signals produced by external sources such as monitoring systems, radio stations, flickering light bulbs, etc. Their position within the image depends on the frequency of the source, the field of view, and the readout bandwidth. Such artifacts are easily eliminated by properly shielding the MR equipment or by removing the source.

5.9 Spurious Echoes Artifacts

A train of several RF pulses in the presence of magnetic field gradients gives rise to many echoes and complicated coherence pathways (Henning, 1991). If other signals besides the desired echo are formed within the sampling window, they interfere adversely and can lead to severe image artifacts. Some of these spurious, unwanted signals are caused by RF pulse errors due to miscalibration or to B_1 inhomogeneities. In a simple spin echo experiment imperfect refocusing pulses will generate spurious FID signals which can last sufficiently long and can be large enough to generate artifacts in the reconstructed image. Such artifacts become more important in pulse sequences with multiple refocusing pulses

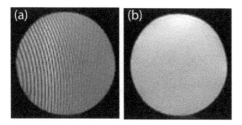

Figure 5.10 Stimulated echo image of a water phantom where (a) no care was taken to remove the unwanted echoes (b) the unwanted echoes were removed by using phase cycling described in Sattin (1985). The image (b) is free of artifacts. Operating frequency 730 MHz. The acquisition parameters are listed in the Appendix.

(FSE, RARE). Even in the case of perfect pulses several echoes will form, depending on the pulse sequence. For example, three consecutive 90° pulses (stimulated echo experiment) will give rise to three free induction decays, three primary spin echoes, one secondary spin echo and one stimulated echo. If only the stimulated echo is intended for creating the image all the other signal contributions must be eliminated to avoid deleterious artifacts (Fig. 5.10).

Phase-cycling schemes for the RF pulses can be used in order to cancel spurious echo signals. A simple example is the use of alternating phases for the refocusing pulses in a spin echo experiment. If two scans with alternating phases are averaged, the FIDs produced by the 180° pulses will cancel while the spin echoes will add coherently. For a stimulated echo experiment a four-step phase cycling applied to the first 90° pulse (0°, 270°, 180°, 90°) and to the receiver (0°, 90°, 180°, 270°) (Sattin, 1985) can be used to obtain artifact free images (Fig. 5.10b). The disadvantage of the phase-cycling approach is that more than one average is necessary to form the image, leading to an increase in the acquisition time.

An alternative strategy for removing spurious signals is to apply spoiling pulsed magnetic field gradients forcing the transverse magnetization to dephase. While this approach will not lengthen the acquisition time, it does present some disadvantages. First, the dephasing induced by magnetic field gradients is position dependent and thus spins close to each other remain mainly in phase. Second, close to the isocenter the dephasing is very small. Third, the insertion of additional gradient pulses within the sequence promotes the formation of eddy currents which can lead to image quality deterioration.

Chapter 6

Sample Preparation

Tissue samples used in MRM are divided in two categories: fixed and alive. Fixed tissues are easier to handle and can withstand long acquisition times. A drawback is that the fixation process can alter the measurements (image SNR and contrast). Alive specimens require perfusion systems adapted to the limited available space and the high magnetic field within the scanner. In this chapter, we describe a number of practical considerations regarding sample preparation and perfusion system design which should be followed in order to ensure good quality MRM images.

6.1 Fixed Tissues

For ex vivo MR measurements, tissue samples are usually chemically fixed, aiming to preserve their in vivo properties as much as possible. Small samples (*Aplysia californica* ganglia, brain slices) can be chemically fixed by immersion in a medium containing a fixation agent. Larger samples (whole mouse or rat brains) are difficult to fix through immersion as they can begin to deteriorate during the time necessary for the penetration of the fixative and before the fixation process is complete. In such cases, it is recommended to perform

Microscopic Magnetic Resonance Imaging: A Practical Perspective
Luisa Ciobanu
Copyright © 2017 Pan Stanford Publishing Pte. Ltd.
ISBN 978-981-4774-71-0 (Paperback), 978-981-4774-42-0 (Hardback), 978-1-315-10732-5 (eBook)
www.panstanford.com

a transcardiac perfusion (perfusion through the left ventricle, for details see Ref. (Dazai, 2011)). The most popular solution used for fixation is 4% formaldehyde in phosphate-buffer solution (PBS), but other fixatives such as 4% glutaraldehyde, or 2% formaldehyde plus 2% glutaraldehyde have also been used. After fixation the samples are typically washed in PBS solution to remove the fixative and then placed in Fluorinert for imaging. The placement of the sample in Fluorinert during imaging is not obligatory but it is recommended as it presents several advantages. First, Fluorinert prevents the sample from drying and, at the same time, does not require a field of view larger than the sample itself as it is proton free (not visible in ^{1}H MRI). In addition, Fluorinert reduces susceptibility artifacts as it has the magnetic susceptibility close to that of cerebrospinal fluid (note that magnetic susceptibility of copper is also similar, which further improves the homogeneity in case of copper coils placed close to the sample, as discussed in Section 2.2.1). Moreover, having the density 1.6 times higher than that of water, Fluorinert can help remove air bubbles trapped within the tissue. Alternatively, the sample can be embedded in an agar gel (Dhenain, 2006).

It is well known that chemical fixation alters tissue characteristics and causes shrinkage. The aldehyde fixatives mentioned previously can significantly and differentially impact several MR parameters. It has been shown that the fixation process reduces both the T_1 and T_2 relaxations times of the tissue. PBS washing prior to imaging has been shown to restore or even prolong the T_2, depending on the fixative, but not the T_1 (Shepherd, 2009). Chemical fixation leads to a reduced SNR in spin density-weighted images which, surprisingly, is not recovered by PBS washing despite the increase in T_2. Diffusion-weighted MR signals are also affected by the fixation process. Specifically, Sun et al. showed that while the fractional anisotropy remains unchanged upon formaldehyde fixation, the apparent diffusion coefficient is significantly reduced (Sun, 2005). Shepherd et al. found significant increased membrane permeability and decreased extracellular space after fixation (Shepherd, 2009).

In addition to the changes in tissue properties, improper fixation can induce severe artifacts rendering the MR images inadequate for quantitative analysis. The concentration of the fixative solution and the timing of the fixation protocol are two key factors. Low aldehyde concentrations ($<2\%$) or short fixation times will lead to tissue deformation and even to the inversion of the white-gray matter T_1 contrast (Cahill, 2012). Long fixation periods (>6 months) lead to neuropil destruction giving rise to severe hypointensities in T_2^*-weighted images of fixed nervous tissue (van Dujin, 2011).

Besides allowing long acquisition times, fixed tissues present the advantage that, after the MR acquisition, they can be histologically examined for correlational studies. However, in the light of the discussion above it is clear that increased attention should be paid to the interpretation of MR images of fixed tissues.

6.2 Live Tissues

MR imaging of live specimens eliminates the fixation issues discussed in the previous section. It brings in interesting opportunities and, at the same time, presents new technical challenges. Live tissue imaging requires the development of dedicated perfusion chambers capable of maintaining its viability and compatible with the strict spatial and material constraints imposed by the high magnetic fields used. Besides the ability to mimic the desired physiological conditions MR compatible perfusion systems should satisfy the following requirements:

(1) All materials should be MR compatible.
(2) The sample should not move during perfusion.
(3) The sample should be fixed using bio-compatible adhesives; Kwik-Sil (World Precision Instruments) is a good choice as it also presents minimal susceptibility artifacts even at very high magnetic fields.
(4) Air bubbles should be eliminated through the insertion of air traps into the system.
(5) The distance between the sample and the RF coil should be kept small. Live perfused specimens are typically imaged using surface coils as the solenoidal geometry is most of the time incompatible with the placement of a perfusion chamber.

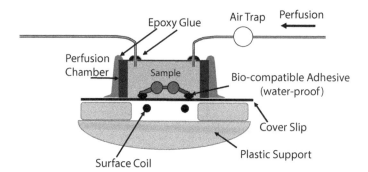

Figure 6.1 Schematic diagram of a simple perfusion system designed for a horizontal bore MR system and surface RF coils. Drawing courtesy of Dr. Yoshihumi Abe.

A schematic of a simple MRM perfusion chamber is shown in Fig. 6.1. This perfusion system was designed for imaging with a surface coil in a horizontal bore magnet but it can be adapted for vertical magnets and for different coil geometries. More sophisticated designs allowing simultaneous analysis of multiple samples have been also proposed (Shepherd, 2002).

SECTION III

APPLICATIONS

Chapter 7

A Bit of History

Magnetic resonance microscopy has been initially defined as magnetic resonance imaging with spatial resolutions on the order of one hundred microns (Glover, 2002; Johnson, 1986). At such resolutions MRM allows the investigation of small animals, mice in particular, with adequate anatomical detail. In this category, performing in vivo longitudinal studies represents one of the main advantages of MRM compared to other imaging techniques. A review of the main MRM applications to live animal imaging is available in Ref. (Badea, 2013). The focus of this book is on magnetic resonance microscopy studies with resolutions between several microns and several tens of microns (which we refer to as high-resolution MRM). Such studies aim at visualizing single cells or small groups of cells and are typically performed on ex vivo or in vitro tissue samples. Recent technological advances made possible the visualization of mammalian neurons; such investigations are, however, very time consuming, preventing dynamic investigations. Systems containing large neurons are definitely advantageous for high-resolution MRM studies. Among these, the marine mollusk *Aplysia* has the largest somatic cells in the animal kingdom. In vertebrates, only eggs can be larger. In the first part of this chapter, Section 7.1, we introduce the *Aplysia* as model system for high-resolution MRM studies, as it

Microscopic Magnetic Resonance Imaging: A Practical Perspective
Luisa Ciobanu
Copyright © 2017 Pan Stanford Publishing Pte. Ltd.
ISBN 978-981-4774-71-0 (Paperback), 978-981-4774-42-0 (Hardback), 978-1-315-10732-5 (eBook)
www.panstanford.com

will be used by the majority of the applications described in the remainder of the book. In the second part we present a brief history of single cell MR microscopy and a survey of recent advances.

7.1 Biological Detour: The *Aplysia*

Aplysia is a marine snail which can be found in subtropical and tropical tide zones throughout the world. There are thirty-seven *Aplysia* species identified, varying in size from a couple of centimeters (*Aplysia parvula*) up to 60–70 cm (*Aplysia giganta*). *Aplysia californica* is a relatively large species (30–40 cm long) found on the California coast (Fig. 7.1a). A comprehensive description of the *Aplysia* can be found in Ref. (Kandel, 1979).

The nervous system of *Aplysia* attracted neurobiologists very early on due to the large size of its neurons. The first electrophysiological studies on *Aplysia*'s neurons were reported by Angelique Arvanitaki in 1940. The model became popular in 1960s when Ladislav Tauc and Eric Kandel started using isolated ganglia to study the cellular mechanism of synaptic palsticity, memory, and learning. Two behaviors often studied in *Aplysia* are the gill withdrawal reflex and the feeding behavior. The gill withdrawal reflex is a behavior in which the animal retracts its gill and siphon as a response to a tactile stimulus. This behavior was found to be sensitive to habituation, sensitization, and classical conditioning. The feeding behavior provides an excellent model system for analyzing and comparing mechanisms underlying appetitive classical conditioning and reward operant conditioning for which behavioral protocols have been developed.

Besides its large nervous cells the *Aplysia* presents other advantages. Its central nervous system resides in five major pairs of bilateral ganglia which are very well separated facilitating therefore the investigation of their specific functions (Fig. 7.1b). The total number of neurons is only ten to twenty thousand, which is also a plus. In addition, the nervous system of *Aplysia* is avascular, meaning that the intact ganglia or the individual neurons can be maintained outside the animal in culture media for long periods of time. Moreover, the ideal temperature for *Aplysia* neurons is

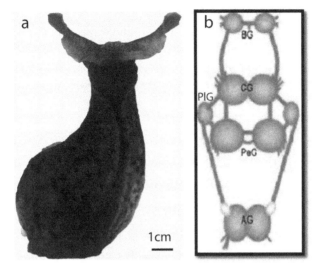

Figure 7.1 (a) Photograph of *Aplysia californica* (b) Schematic of *Aplysia*'s nervous system. BG = Buccal Ganglia, CG = Cerebral Ganglia, PeG = Pedal Ganglia, PlG = Pleural Ganglia, AG = Abdominal Ganglion.

between 15 and 25°C which drastically simplifies their manipulation compared to that of mammalian neurons.

Each ganglion has three distinct components: a surrounding connective tissue sheath, a peripheral region consisting of cell bodies and a central region (neuropil) containing axons and dendrites. The sheath has a structural role enclosing not only the ganglion but also the connective nerves (fiber tracts) between the ganglia. This tissue is permeable to ions but impermeable to large molecules. Due to a membrane bound carotenoid pigment the somas of *Aplysia* neurons are bright yellow or orange. A difference between vertebrate neurons and *Aplysia* neurons is that the latter form synapses only on the dendritic arbor of the main axon and never on the cell body. The cytoplasm of an *Aplysia* neuron contains the typical components found in vertebrates: mitochondria, endoplasmatic reticulum, ribosomes, microtubules, Golgi apparatus, neurofilaments and vesicles. The nuclei are round or oval and occupy approximately two thirds of the cell volume. The nuclei of large neurons typically contain thousands nucleoli. As most invertebrates *Aplysia* contains also different types of glial cells located in the neuropil, cell body-layer and connectivities. While we have a lot of information about *Aplysia*'s neurons, little is known about the glial cells.

Among the five pairs of ganglia shown in Fig. 7.1 the ones typically used in MRM studies are the buccal and abdominal ganglia, to be described in what follows.

7.1.1 The Buccal Ganglia

The buccal ganglia are the smallest of the five ganglia (volume-wise) and they contain many large neurons ranging in diameters from 100 to 200 μm. Located towards the head end of the animal the buccal ganglia innervates the muscles of the buccal mass controlling protraction and retraction of the radula (a tongue-like organ) and the motility of the esophagus, the pharynx and the salivary glands. In each ganglion more than 50 cells that are responsible for generating the radula movements and several clusters of sensory neurons have been identified (Fig. 7.2). The identified neurons have been labeled with the letter B (Buccal) and a number (B1, B2, B3, etc.) generally in chronological order of their identification, which is essentially the decreasing order of their size. The buccal ganglia of *Aplysia* is ideal for the study of the functional properties of a central neuronal network generating a motivated behavior and its plasticity induced by non-associative and associative learning (Brembs, 2002; Kupfermann, 1974; Nargeot, 1997).

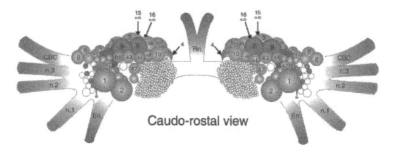

Figure 7.2 Schematic of the buccal ganglia with nerves and most of the big neurons labeled. Drawing courtesy of Dr. Romuald Nargeot.

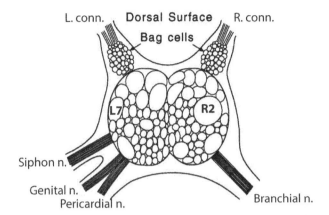

Figure 7.3 Schematic of the abdominal ganglion showing the main nerves and the positions of neurons R2 and L7. Adapted from Kupfermann (1974).

7.1.2 The Abdominal Ganglion

The abdominal ganglion, also known as the visceral ganglion, is located near the anterior aorta. Unlike the other ganglia, the abdominal ganglion is asymmetric. Most of the large neurons within this ganglion have been identified and labeled L and R (for left and right hemiganglion, respectively) and assigned a number (Fig. 7.3). The first successful cellular analyses of learning were performed on this ganglion (for which Eric Kandel was awarded the 2000 Nobel Prize in Medicine). The giant neuron R2 is the largest cell in *Aplysia*'s nervous system: it can reach 1 mm in size with a 500 μm diameter nucleus for adult animals. Neurons R2 and L7 are often used in single cell MRM studies.

7.2 Advances in Spatial Resolution

In 1986 Aguayo et al. demonstrated that by combining the advantage of strong magnetic fields (9.4 T) and small RF coils (∼5 mm diameter) it is possible to boost the sensitivity of MR experiments and significantly improve the spatial resolution, which until that date was only on the order of millimeters (Aguayo, 1986). Shown in Figs. 7.4a and 7.4b are images of ova obtained from *Xenopus*

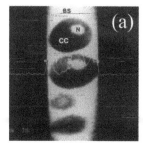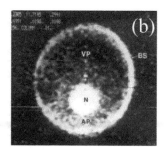

Figure 7.4 (a) Proton MR image at 9.5 T of four ova from *Xenopus laevis* at different stages of oogenesis. The tube containing the ova has inner diameter 1.1 μm, and runs vertically in the figure. The thickness of the slice shown is 500 μm. The in-plane image matrix is 256 × 256, with resolution 16 × 27 μm². (b) Proton MR image of a transverse slice (slice thickness 250 μm) across a glass tube (inner diameter 1.1 mm) containing a stage-4 *Xenopus laevis* ovum bathed in Barth's solution. Reprinted by permission from Macmillan Publishers Ltd: [Nature] (Aguayo, 1986), copyright (1986).

laevis. The sample, containing cells with diameters on the order of 1 mm, was placed inside a 1.1 mm ID glass capillary. These earliest single cell images, with resolutions of 10 μm × 13 μm × 250 μm already display some of the rich image contrast mechanisms of MR microscopy. In Fig. 7.4b the cell cytoplasm appears nearly black relative to the free water outside the cell, despite the fact that the proton densities in these two regions are not totally dissimilar. The nucleus is clearly observed, with brightness similar to that of the external water. The contrast then must result from differing relaxation (T_1 or T_2) behaviors of the cytoplasm and external water, and the particular imaging pulse sequence and data acquisition parameters.

An abstract from 1989 by Zhou et al. (1989) reported fully three-dimensional MR images with spatial resolution of $(6.37 \ \mu m)^3$ or 260 μm^3 (260 fl) however, the images themselves have not been published. While better in-plane resolutions have since been obtained (Bowtell, 1990; Cho, 1990, 1988), the volume resolution of Zhou et al. was not exceeded until 2001 by Lee et al. who report 2D images having volume resolution 75 μm^3 and in-plane resolution of 1 μm × 1 μm (Lee, 2001). Fig. 7.5 shows the image of a

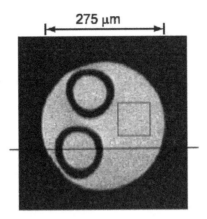

275 μm

Figure 7.5 Proton MR image at 14.1 T of a tube containing two capillaries with voxel size of $1 \times 1 \times 75$ μm^3. The diameter and thickness of two capillaries are about 110 and 16 μm, respectively. Reprinted from Lee (2001) with permission from Elsevier.

phantom consisting of a cylinder filled with hydrocarbon composite oil in which two capillaries with 110 μm outer diameter have been inserted. The image was obtained at 14.1 T using a 500 μm diameter RF micro-coil and gradients of about 10 T/m. The same experimental setup was used for in vivo imaging of a geranium leaf steam with 2 μm in plane resolution, for a 200 μm^3 voxel volume.

While the images obtained by Lee et al. have excellent in-plane resolution, the relatively large slice thickness is not adequate for imaging objects lacking 2D symmetry. By employing magnetic field gradients as large as 5.8 T/m, micro-receiver coils with diameters smaller than 100 μm Ciobanu et al. report fully three-dimensional images obtained on both phantoms and real biological samples. Fig. 7.6 shows the 3D image from Ref. (Ciobanu, 2002), a quartz microcapillary, initially 1 mm outer diameter, pulled to an outer diameter of ~73 μm and inner diameter ~53 μm, filled with water and 39 μm diameter fluorescent polymer beads. The 3D image is presented in 25 successive xz plane panels (with x being the axis of the micro-pipette and z the direction of the applied field). Ref. (Ciobanu, 2002) provides a quantitative assessment of the resolution of this image; the resolution along the x, y and z axes was 3.7 μm $\times$ 3.3 μm $\times$ 3.3 μm, respectively.

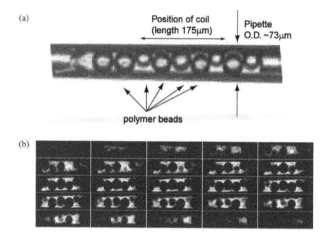

Figure 7.6 (a) Microscope photograph of sample imaged in (b). (b) 3D MR microscopy image of the sample shown in (a) (acquired at 9 T). Defining the x direction as the axis of the pipette and the direction z parallel to the applied field, the resolution along the x (y) [z] axis is 3.7 μm (3.3 μm) [3.3 μm] for a field of view of 237 μm (66 μm) [66 μm]. Reprinted from Ciobanu (2002) with permission from Elsevier.

Figure 7.7 shows the Ciobanu et al. image of a spirogyra alga (Ciobanu, 2003). The spirogyra (see microscope photo in Fig. 7.7a) is cigar-shaped, with diameter ~40 μm. The micropipette containing the spirogyra cell, immersed in water, as shown in Fig. 7.7a, has inner diameter ~55 μm. The microcoil, 250 μm long, covers 5–6 chloroplast spiral windings. The MR image is shown in Fig. 7.7b. The panels 4–9 in the image have clear arrays of 5–6 black spots near both the top and the bottom of the panel. These black spots correspond to the expected 5–6 piercings of the image plane by the spiral chloroplast array.

Using surface RF microcoils, fast switching magnetic field gradients and very high static magnetic fields (18.8 T) Weiger et al. (2008) obtained images of a fleece of glass fibers immersed in water with 3 μm isotropic resolution, which is today the highest resolution reported for MRM. Recently, Lee at al. reported MRM of *Aplysia* L7 neurons with resolutions of $7.8 \times 7.8 \times 15$ μm^3 (Lee, 2015). At this resolution the authors were able to identify subcellular structures

(a)

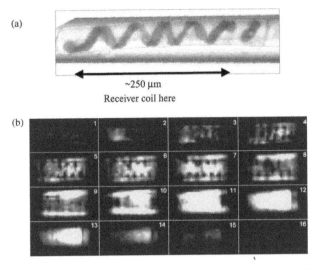

~250 μm

Receiver coil here

(b)

Figure 7.7 (a) Microscope photograph of the sample imaged in (b). The sample consists of a single-cell spirogyra alga of cylindrical shape with diameter ~40 μm and length of several hundred microns. The NMR receiver coil is wound over a span of ~250 μm. (b) 3D MR microscopy image of the sample shown in (a) (acquired at 9 T). Reprinted from Ciobanu (2003) with permission from Elsevier.

such as nucleus, cytoplasm but also nuclear and plasma membranes which have been correlated with histological images.

The possibility of imaging mammalian neurons has also been demonstrated. Flint et al. report MRM images of chemical fixed rat striatum with 4.7 μm isotropic resolution (Flint, 2009a). The same authors obtained images of human, fixed spinal cord tissue in which cell bodies and neural processes of α-motor neurons can be visualized (Flint, 2012) (Fig. 7.8). In both studies the authors used a magnetic field of 11.7 T and custom microimaging gradients with maximum strength of 3 T/m and a 500 μm surface microcoil.

Thus, since 1986 MR microscopy has advanced to voxel resolutions of just a few microns in all three spatial dimensions, and to as little as one micron "in-plane" resolution. Such studies are, however, by no means straightforward to perform and demand specialized equipment and optimized methods. Moreover, as it can be seen in

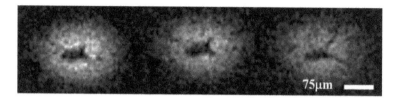

Figure 7.8 Representative MRM images (6.25 μm isotropic resolution) taken from three-dimensional T_2-weighted datasets illustrating the hypointense projections emanating from cell bodies of α-motor neurons in human spinal cord. Reprinted from Flint (2012) with permission from Elsevier.

Table 7.1 Some of the highest resolution MRM images reported to date and the required acquisition time

Reference	Sample	Spatial Resolution (μm^3)	Acquisition Time (hours)
Ciobanu (2002)	Polymer beads in water	$3.7 \times 3.3 \times 3.3$	30
Weiger (2008)	Glass fibers in water	3.0^3	58
Flint (2009a)	Chemically fixed rat striatum	4.7^3	22
Flint (2012)	Chemically fixed human spinal cord	6.25^3	64

Table 7.1, they are time-consuming, typically requiring tens of hours of data collection, preventing, in most cases, the investigation of living tissues.

While MR microscopy cannot compete with many traditional imaging methods in terms of spatial resolution (for example, light microscopy with resolution of hundreds of nanometer or electron microscopy at tens of angstroms), it compensates by providing a rich array of contrast variables. Many of these, as already discussed in previous chapters, are widely known and used in more traditional, lower resolution MRI and include relaxation times T_1 and T_2 (and notably T_2^*; with its role in functional MRI), and the presence of flow and diffusion. In the following chapters we will discuss the use of some of these different contrast mechanisms in high-resolution imaging of biological tissue. The majority of studies to be presented are performed on systems with large neurons, such as the *Aplysia* ganglia, as they allow the visualization of single neurons or even of subcelullar structures in living samples. We point out that the goal of resolving single mammalian neurons in living tissues has not been abandoned and it continues to represent an active area of research.

Chapter 8

Diffusion Weighted Magnetic Resonance Microscopy

Since its introduction in the mid-1980s, diffusion-weighted magnetic resonance imaging has become one of the main tools of modern neuroimaging studies. At small scales, magnetic resonance microscopy investigations have an important role in understanding how the diffusion MR signal is related to tissue microstructure and in developing realistic models. As we will see in the second part of this chapter, recent evidence shows that changes in water diffusion can reflect neuronal activity. Therefore, by providing data at microscopic scales, diffusion MR microscopy has the potential uncover the biophysical mechanisms underlying neuronal activation.

8.1 Diffusion and Tissue Microstructure

Unlike the diffusion signal decay in a homogeneous medium, the diffusion signal behavior in tissues is not monoexponential. Instead, biexponential functions where empirically found to provide good fits. The initial interpretation was that the two exponentials correspond to the *intra* and the *extra* cellular components. This

Microscopic Magnetic Resonance Imaging: A Practical Perspective
Luisa Ciobanu
Copyright © 2017 Pan Stanford Publishing Pte. Ltd.
ISBN 978-981-4774-71-0 (Paperback), 978-981-4774-42-0 (Hardback), 978-1-315-10732-5 (eBook)
www.panstanford.com

hypothesis was, however, unsubstantiated by the first diffusion measurements at single cell level reported by Schoeniger et al. (1994). The authors studied *Aplysia californica* neurons and showed that two different diffusion components exist within one cell: one relatively fast ($ADC_{fast} \sim 1.46$ μm^2/ms) within the cell nucleus and one slower ($ADC_{slow} \sim 0.279$ μm^2/ms) within the cytoplasm. Hsu et al. (1996) also found smaller ADC values in the cytoplasm compared to the nucleus, but overall they report values smaller than those measured by Schoeniger. In a recent study Jelescu et al. (2014) found mean ADCs values in the cell, cytoplasm and nucleus of (0.68 ± 0.04), (0.57 ± 0.04) and (0.91 ± 0.05) μm^2/ms, respectively, in good agreement with the values reported by Hsu et al. In addition, using the same animal model, Grant and co-authors (2001) concluded that, while water diffusion is practically monoexponential in the nucleus, a non-monoexponential behavior is observed in the cytoplasm.

Studies performed on other large cells also confirmed that the presence of the extracellular space is not necessary in order to observe the multi-exponential diffusion behavior. In a series of papers, Sehy et al. investigated the diffusion properties of the stage 5 *Xenopus laevis* oocyte (Sehy, 2001, 2002a,b). The first of these studies reports ADC values of 1.7 μm^2/ms in the nucleus while the ranges found in the cytoplasm were from 0.86 μm^2/ms in the animal pole to 0.57 μm^2/ms in the vegetal pole (Sehy, 2001).

In a second paper (Sehy, 2002b), the authors measured the intracellular water ADC by doping the extracellular medium with a contrast agent which reduced the T_1 relaxation time: spectroscopic diffusion measurements were performed, rather than imaging, in order to reduce the influence of lipid signals. The results showed, once again, that water diffusion inside the cell is biexponential with 89% having a diffusion coefficient of 1.06 ± 0.05 μm^2/ms and 11% a value of 0.16 ± 0.02 μm^2/ms. Hyperosmolar medium was then added to the perfusate such that the cells volume increased by $16 \pm 4\%$, as measured by 3D MRI. The ADCs of the fast and slow components both increased, to 1.27 ± 0.03 and 0.22 ± 0.02 μm^2/ms, respectively, but the volume fractions remained constant. In a third paper (Sehy, 2002c), the authors investigated the diffusional membrane permeability. A model was proposed relating intracellular lifetime to true membrane diffusional permeability. The latter quantity was measured to have a value of 2.7 ± 0.4 $\mu m^2/ms$, which is approximately 40% greater than the apparent diffusional permeability.

Tissue geometrical parameters (size of neurons and axons, myelin thickness, neurite orientation distribution) can be obtained by comparing numerical simulations of simple tissue models with experimental data. Numerous biophysical models have been proposed, most of them assuming that the tissue is composed of an extracellular space and spherical, ellipsoidal or cylindrical cellular components. To validate these models, experiments have been conducted using artificial phantoms: straight polyester crossing fibers (Pullens, 2010), crossing fibers wrapped on spherical spindles (Moussavi, 2011), resected rat spinal cords (Campbell, 2005), and carrot slices (Dietrich, 2014).

The majority of diffusion models employed currently make the assumption that the tissue under investigation is dominated by a single compartment with a typical structure size. The presence of the nucleus within the cells is not taken into consideration. However, recent results comparing theoretical predictions and experimental data highlight the importance of including a nucleus with a different intrinsic diffusivity from that of the cytoplasm when modeling cells within tissues. ADC measurements were performed in 22 big *Aplysia* neurons (radii between 100 and 200 μm) for five diffusion times: $\Delta = 5, 10, 15, 20$ and 25 ms. From these measurements, we fit the ADC as $A\sqrt{\Delta} + B$ (see Eq. 4.18) and thus extract the coefficients A and B. Considering the cells homogeneous spheres and knowing that $A = -D_0 \dfrac{4\sqrt{D_0}}{3\sqrt{\pi}} \dfrac{1}{R}$ and $B = D_0$, an estimated cell radius, R_{est}, can be calculated as:

$$R_{est} = -\frac{4}{3\sqrt{\pi}} A^{-1} B^{3/2}, \tag{8.1}$$

where D_0 is the intrinsic diffusivity. For the 22 cells considered in this study the estimated cells' radii were always much smaller than the actual sizes, measured on high-resolution T_2-weighted images (77% average difference) (Nguyen, 2017).

The cause of this discrepancy was investigated by performing numerical simulations of ADCs within cells consisting of two compartments, cytoplasm and nucleus, for different diffusion encoding times Δ. The results showed that the dependence of ADC on Δ is

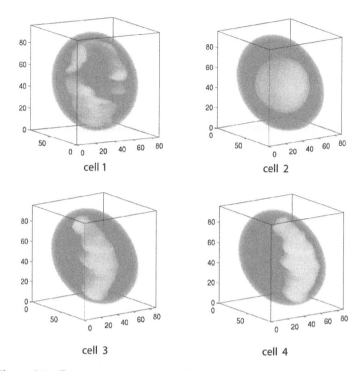

Figure 8.1 Four computer generated cells. The volume fraction of the nucleus in all four domains is 25%.

different for cells with a nucleus compared to those without, that it is impacted by the nuclear volume fraction and moreover, that the shape of the nucleus has a very strong influence. Specifically, four different cell geometries were considered. Among these, three contain different shape nuclei (Fig. 8.1). Cell 3 and cell 4 have the same shape of the nucleus but its placement is different inside the cell. In all four cases, the cells have the same size, shape and nuclear volume fraction. The ADCs were calculated numerically by solving the Bloch–Torrey equation (Eq. 4.11) for different cells radii (70–500 µm) and diffusion times (5–25 ms) considering intrinsic diffusivities in the cytoplasm and nucleus of 1 and 2 µm²/ms, respectively. As previously, the slope A was calculated from the dependence of ADC on Δ. As illustrated in Fig. 8.2, for all radii considered, A strongly depends on the nucleus shape (Nguyen, 2017) but it does not depend on its position within the cell. Knowing

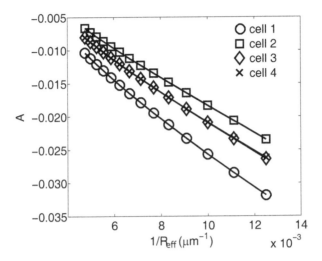

Figure 8.2 The dependence of slope A on the effective cell radius R_{eff} for the four geometries shown in Fig. 8.1. A strongly depends on the shape of the nucleus but not on the position of the nucleus within the cell (cell 3 and cell 4 produce the same results).

that the nucleus shape is closely linked to cellular function (Webster, 2009), this implies that ADC measurements can serve as biomarkers in various pathologies.

Despite their large size, the giant cells used in the studies presented above appear to accurately reflect the behavior of water diffusion in smaller mammalian neurons. However, relating tissue microarchitecture characteristics to the macroscopic MR diffusion signal remains a difficult task. The development of numerical simulations of more complex and realistic models is necessary, and magnetic resonance microscopy will continue to play an important role in their validation.

8.2 Diffusion and Neuronal Function

Another application of diffusion MRI is the measurement of neuronal activity. Diffusion based functional MRI (DfMRI) studies have been reported in human subjects (Le Bihan, 2006) and animal models (Tsurugizawa, 2013). The hypothesis behind DfMRI is that transient neuronal network morphological changes (e.g., cell swelling) accompanying neuronal activity lead to a detectable decrease in the apparent diffusion coefficient of the tissue. As of today, there is no general consensus on whether this hypothesis is true, and the precise origin of the DfMRI signal is still unclear. MRM investigations hold the potential to identify the morphological changes occurring at the cellular level during neuronal activation and establish whether or not they are correlated to the detected diffusion MR signal changes. Such studies, typically performed on tissue samples, present several advantages. First, they avoid confounding factors such as blood flow, blood oxygenation changes, motion artifacts related to breathing or anesthesia effects. Second, they significantly simplify result interpretation by reducing the complexity of the networks investigated.

Experiments performed on live hippocampal slices demonstrated diffusion changes induced by kainate or potassium (Flint, 2009b). However, the spatial resolution at which these experiments were performed (156×156 µm in plane) did not allow the identification of specific hippocampal subregions with increased neuronal activity.

Using the *Aplysia* buccal ganglia as model system, water diffusion has been measured inside the soma of single neurons and in the region of cell bodies upon exposure to ouabain, which inhibits Na^+/K^+ pumps (Jelescu, 2014). The results showed an increase in water diffusion inside isolated neurons but a decrease at tissue level (Fig. 8.3). Similarly, the application of dopamine to the buccal ganglia of *Aplysia californica* led to an ADC decrease in the soma but and increase at ganglia level (Abe, 2017). In both studies, cell size measurements, performed either with MRM or with optical microscopy, revealed neuronal swelling. Moreover, the dopamine-induced cellular ADC increase in individual neurons was found to be significantly correlated with the increase in the cell diameter and the cell volume.

These results can be explained considering the existence of a layer of water molecules bound to the cell membrane surface presenting a smaller apparent diffusion coefficient as proposed by Le Bihan (2014). Such a layer would increase upon cells

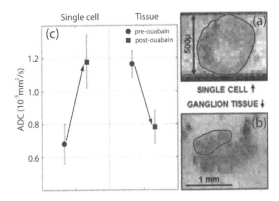

Figure 8.3 Diffusion measurements in a single neuron and in buccal ganglia of *Aplysia californica*. (a) Single-neuron MR image with 25 μm isotropic resolution; (b) Buccal ganglia image showing the region for ADC measurements; (c) ADC measurements pre and post ouabain treatment. The values obtained pre and post ouabain treatment are statistically significant for both single cell and tissue experiments. Adapted from Jelescu et al. (2014) with permission from John Wiley and Sons.

swelling decreasing thereby the overall ADC in the tissue. Higher spatial resolution is necessary in order to evaluate ADC changes in subcellular compartments and to elucidate whether indeed the ADC near cell membranes is lower than it is within the cytoplasm.

Chapter 9

Manganese Enhanced Magnetic Resonance Microscopy

In the last decade, a new functional MR technique, manganese-enhanced MRI (MEMRI), has been successfully proven on various vertebrate animal models (Pautler, 2002; Silva, 2004; Van der Linden, 2004). MEMRI uses manganese, an MR contrast agent, to label active neurons.[a] The amount of manganese that accumulates intracellularly is directly linked to neuronal activity, because the Mn^{2+} ions enter the neurons through calcium and nonspecific cationic channels and can be transported along axons and across synapses (Crossgrove, 2005; Geiger, 2009; Narita, 1990; Nelson, 1986). Once accumulated inside the neurons, manganese can be detected by acquiring T_1 relaxation maps or simply by measuring the signal intensity in T_1-weighted images (water protons located in the vicinity of Mn^{2+} ions will have shorter relaxation times).

There are several advantages of MEMRI over traditional fMRI techniques such as Blood Oxygen Level Dependent (BOLD). When performing functional MEMRI experiments, the stimuli are presented before the imaging session and therefore the confounding

---

[a]When used to measure neuronal activity, the technique is sometimes referred to as activation-induced MEMRI (AIM-MRI) (Lin, 1997).

Microscopic Magnetic Resonance Imaging: A Practical Perspective
Luisa Ciobanu
Copyright © 2017 Pan Stanford Publishing Pte. Ltd.
ISBN 978-981-4774-71-0 (Paperback), 978-981-4774-42-0 (Hardback), 978-1-315-10732-5 (eBook)
www.panstanford.com

effects related to the anesthesia are avoided. For studies performed on rodents, the animals are injected with a $MnCl_2$ solution 24 h before performing the MRI. The stimulus is typically presented to the animals within this 24 h window. Once inside neuronal circuits; the Mn^{2+} ions are eliminated very slowly, and this allows for longer acquisition times improving image spatial resolution and SNR. There is, however, one drawback: the signal measured reflects the integrated activity over a long period of time; rapid changes, and especially deactivation, cannot be detected.

MEMRI studies performed in vertebrates reach spatial resolutions of 100 μm in all three directions. Even at these resolutions one can only visualize brain regions composed of clusters of hundreds of neurons. Manganese-enhanced magnetic resonance microscopy aims to detect neuronal activation in single cells. In what follows we will divide the microscopy MEMRI experiments into two categories. In the first category the $MnCl_2$ administration and the stimulation are performed in vivo, in intact, freely behaving animals, while in the second category they are performed ex vivo, on live tissue/organ specimens. In both cases the imaging is performed on isolated, ex vivo tissue/organ samples.

9.1 In vivo Manganese Administration

Herberholz et al. (2004, 2011) extended the use of MEMRI to the study of invertebrate animal models. The authors succeeded in labeling activity-dependent Mn^{2+} uptake in the nervous system of crayfish and produced activation maps with 78 μm isotropic resolution. At this resolution it was not possible to identify individual neurons, complicating the interpretation of their results.

The first functional MRI studies with single-neuron resolution were reported by Radecki et al. (2014) using as model system the marine mollusk *Aplysia californica*. The experimental protocol for manganese administration in rodents has been adapted to the *Aplysia*. Specifically, it is has been found that the optimum time for imaging was between 45 and 90 min after the $MnCl_2$ injection. This short time compared to vertebrate protocols is due to the open circulatory system found in this animal species (see Section 7.1.1). After injection with MnCl$_2$ solution the animals were allowed to freely move in the aquarium while being presented different stimuli. For imaging, the ganglion of interest was dissected (buccal ganglia for food stimuli) and placed in a capillary filled with artificial sea water. For each pair of ganglia, two images were acquired: a 3D FLASH for T_1 contrast and a 3D RARE for T_2 contrast. The cell bodies display a natural T_2 contrast with respect to artificial sea water, making them visible on T_2-weighted images regardless of whether they accumulated Mn^{2+} or not. The T_1-weighted images reveal regions of Mn^{2+} uptake only, that is no neurons can be identified on FLASH images of ganglia coming from an animal which was not injected with MnCl$_2$ solution. MEMRI signal intensities were measured in individual neurons. As in vertebrates, it was found that Mn^{2+} accumulates intracellularly when injected into the living animal and that the intracellular Mn^{2+} concentration increases in response to sensory stimuli (Fig. 9.1).

In the same study, Radecki et al. showed that it is possible to distinguish between different neuronal responses induced by

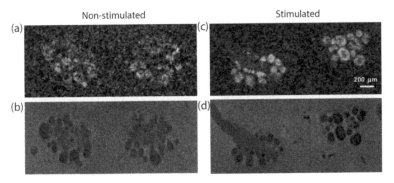

Figure 9.1 Representative MR images of ganglia from nonstimulated (left) and stimulated (right) animals. Selected slices from T_1-weighted images showing intracellular Mn^{2+} accumulation are displayed in panels (a) and (c). The hyperintense regions represent individual neurons that have been identified using the T_2-weighted images shown in (b) and (d). Spatial resolution: 25 μm isotropic. Operating frequency 730 MHz. (Scale bar: 200 μm.) Copyright (2014) National Academy of Sciences.

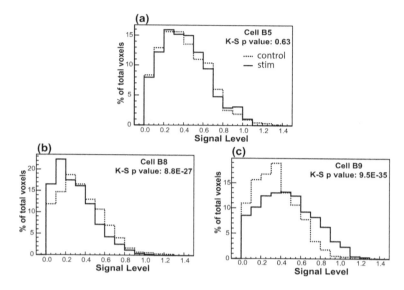

Figure 9.2 Distributions of voxel signal intensities (percentage of all voxels) obtained for neurons B5 (a), B8 (b), and B9 (c) for control (continuous line) and arousing stimuli (dotted line). The distributions corresponding to the two conditions are statistically different for neurons B8 and B9, but not for neuron B5. For neuron B8, the histogram corresponding to the stimulated state is shifted to the left compared with the control, suggesting a decrease in neuronal activity. For neuron B9, the situation is reversed, indicating that neuronal activity increases upon stimulation. Copyright (2014) National Academy of Sciences.

a particular stimulus. In addition to the increase in the Mn^{2+} concentration as observed in most neurons within the buccal ganglia, no change or a concentration diminution was revealed in specific neurons (neurons B5 and B8, respectively in Fig. 9.2) upon presenting arousing only stimuli (not followed by a reward). Therefore, these particular stimuli seem to have no effect on neuron B5 and to decrease or inhibit the spontaneous electrical activity of neuron B8.

The absolute quantification of the Mn^{2+} concentration accumulated intracellularly is possible by acquiring T_1 relaxation maps. Due to their time consuming nature such acquisitions were performed at lower spatial resolution ($30 \times 30 \times 100$ μm^3 as opposed to

$25 \times 25 \times 25 \ \mu m^3$ used for the FLASH and RARE images described previously). As a result, only the biggest neurons (B1, B2, B3, and B4, see Fig. 7.2 in Section 7.1.1) could be investigated with minimum partial volume effects. Within these neurons, the Mn^{2+} concentration computed based on the measured T_1 relaxation times confirmed the dependence on the sensory stimulation as revealed by the FLASH images.

The method implemented by Radecki et al. can be extended to other networks within the *Aplysia* nervous system such as the cerebral ganglion, which is responsible for overall control and coordination of behavior, or the pedal-pleural ganglia, which mediate locomotion and head-weaving behavior.

Despite its simplicity, the nervous system of *Aplysia* can constitute a good model for studying age-related conditions or specific diseases. Electrophysiological experiments showed that aging perturbs the neuronal circuit for tail withdrawal reflex resulting in learning failure (Kempsell, 2015). As early as 1983, E. Kandel proposed that both anticipatory and chronic anxiety disorders can be modeled in the *Aplysia* (Kandel, 1983). Reward-induced compulsive behaviors in vertebrates bear strikingly similar features with the changes in the feeding behavior of *Aplysia* induced by operant conditioning (Nargeot, 2007). These examples represent just a few areas in which high-resolution MEMRI studies can be used to characterize the associated neuronal activity and plasticity changes.

One can also envision using the same procedure to investigate vertebrate nervous systems with single cell resolution. As in the case of the *Aplysia*, non-anesthetized animals would be injected with $MnCl_2$ and presented with specific stimuli. Subsequently, the tissues of interest would be excised for high-resolution imaging. Chemical fixation of this tissue can be used in order to allow longer acquisition times. Special care should be taken as the Mn^{2+} distribution may change in time due to structural and chemical changes induced by fixation (Liu, 2013). Given the much smaller size of the mammalian neurons, such studies will be more challenging and will likely require new, innovative designs for both the RF and the gradient coils.

9.2 Ex vivo Manganese Administration

High-resolution MEMRI investigations can also be carried out by administering the Mn^{2+} to ex vivo specimens, with the advantage that they can be easily combined with other techniques such as electrophysiology and optical microscopy. As certain behaviors can be successfully mimicked in vitro (the operant conditioning in *Aplysia*, for example) functional studies can be performed using this approach. Such experiments are conducted by bathing the specimens into $MnCl_2$ solution while presenting a chemical or electrical stimulus.

Stimulating the *Aplysia* buccal ganglia with the neurotransmitter dopamine leads to higher intracellular manganese concentrations in all neurons compared to the baseline state. Similarly, images of buccal ganglia acquired after electrically stimulating the esophageal nerve for 45 min present significantly higher MEMRI signal intensities than those of nonstimulated ganglia (Fig. 9.3).

In addition to their application to functional studies, ex vivo MEMRI experiments can help elucidate the exact mechanism through which the Mn^{2+} ions enters neurons and gets transported within the nervous system. Jelescu et al. confirmed the retrograde Mn^{2+} transport along one single nerve (Jelescu, 2013). Specifically, by dipping one nerve of the excised buccal ganglia in $MnCl_2$ solution the authors report images in which all ipsilateral motor neurons with known axonal projections into that specific nerve (previously identified with Co^{2+} or Ni^{2+} migration and optical techniques) show as hyperintense in T_1-weighted images (Fig. 9.4). The same study showed that chemical stimulation with dopamine alters the dynamics of Mn^{2+} inside the ganglia, namely that the natural washout in time becomes dominated by Mn^{2+} transfer from initially loaded cells to unloaded cells.

Comparisons between MEMRI and electrophysiological measurements support the hypothesis that action potential firing plays an essential role in the intracellular Mn^{2+} accumulation and, additionally, suggest that subthreshold depolarization can also lead to Mn^{2+} influx in non-firing neurons, although to a

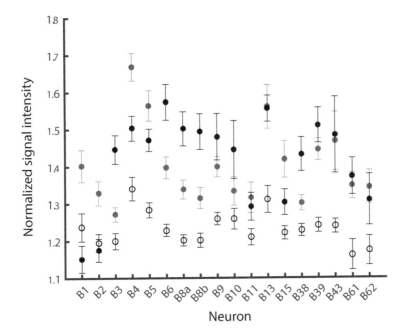

Figure 9.3 Normalized MEMRI signal intensity in individual neurons for control (○), dopamine stimulation (50 μM) (●) and electrical stimulation (2 Hz, 8 V, 0.3 ms pulse) (●). Each point is obtained by averaging the signal in 10 neurons (5 animals) and the error bars represent the SEM.

smaller extent (Svehla, 2017). Additional studies are necessary in order to understand exactly how Mn^{2+} accumulates intracellularly as well as to characterize its dynamics within the nervous system. While it is reasonable to assume that the *Aplysia* will help accomplish this goal, extra attention should be paid when transferring the results to vertebrates given the differences that exist between the calcium channels present in the two animal classes (Byerly, 1985).

9.3 Manganese Toxicity

The main drawback of using MEMRI for functional studies is the toxicity of manganese. In humans, overexposure to manganese leads

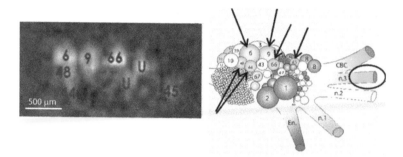

Figure 9.4 Schematic of the right side of the buccal ganglia (right). MR image showing the accumulation of Mn^{2+} following retrograde transport along nerve n.3 (left). The enhanced cells on the T_1-weighted image are those with axonal projections in n.3 (indicated with black arrows in the schematic). U = unidentified cell. Figure courtesy of Dr. Ileana Jelescu.

to manganism, a psychiatric disorder characterized by Parkinson's-like symptoms (Mena, 1967). At cellular level, at concentrations higher than 250 μM, manganese is a neurological and cardiac toxin as demonstrated by studies in rodents (Aschner, 1994; Brurok, 1997). Jelescu et al. (2013) showed that intracellular manganese concentrations on the order of 250 μM did not alter the bioelectrical properties of the neuronal network and the neurons in the *Aplysia*. We note that the intracellular manganese concentrations reported by Radecki et al. in reference (Radecki, 2014) and Table 9.1 are all lower than 250 μM. Different species and even different types of neurons may have different toxicity thresholds and therefore a careful assessment of the maximum usable manganese dose is

Table 9.1 Mn^{2+} concentrations in neurons within the buccal ganglia of *Aplysia* for non-stimulated and stimulated conditions. The errors represent SEM ($n = 10$). Copyright (2014) National Academy of Sciences

Neuron	Intracellular Mn^{2+} concentration (μM)	
	Non-stimulated	Stimulated
B1	18.3 ± 7.7	53.1 ± 22.1
B2	17.6 ± 7.0	51.6 ± 16.5
B3	22.6 ± 9.0	84.1 ± 29.3
B4	26.3 ± 10.0	64.4 ± 26.1

necessary for each particular case. As a general rule, in all MEMRI experiments the MR sequence parameters should be optimized in order to improve the sensitivity and enable the detection of small concentrations of Mn^{2+}, reducing therefore the risk of toxicity.

SECTION **IV**

CONCLUSION

Chapter 10

On the Horizon

The full potential of magnetic resonance microscopy has yet to be achieved. Higher magnetic fields and advances in microcoil and circuitry designs will lead to further improvements in sensitivity and spatial resolution. We will most certainly see a growth in the use of cryogenic probes facilitated by progress in cooling technologies, allowing enhanced thermal stability and insulation. It is conceivable that much higher static magnetic fields will be made available through the development of pulsed field magnets. In parallel, the development of increasingly more sophisticated data acquisition and reconstruction methods will significantly reduce the image acquisition times.

All these technological developments will considerably widen the range of MRM applications. Biological systems containing large cells such as the *Aplysia* will continue to be investigated, and the scope of such investigations will be enlarged. As an example, these could include the assessment of anatomical, functional, and molecular modifications induced by behavioral changes. Anatomical imaging of fixed, isolated mammalian brain tissue with cellular and subcellular resolutions is already a reality. The possibility of imaging and even performing functional studies of single mammalian neurons is within reach.

Microscopic Magnetic Resonance Imaging: A Practical Perspective
Luisa Ciobanu
Copyright © 2017 Pan Stanford Publishing Pte. Ltd.
ISBN 978-981-4774-71-0 (Paperback), 978-981-4774-42-0 (Hardback), 978-1-315-10732-5 (eBook)
www.panstanford.com

Appendix

This Appendix summarizes the acquisition parameters for the images acquired specifically for this book or those not previously published. All acquisitions were performed using a 17.2 T small animal imaging system. The following abbreviations are used:

TE	echo time
TR	repetition time
TI	inversion time
FOV	field-of-view
thk	slice thickness
BW	bandwidth
NA	number of averages
AF	acceleration factor
Nseg	number of segments
RG	receiver gain
DW	diffusion weighted
DP	diffusion prepared
δ	diffusion gradient duration
Δ	diffusion gradient separation

Table A.1   Fig. 2.3

Pulse sequence	2D RARE
TE/TR (ms)	12/3000
FOV (mm)	12.8 × 6.4
thk (mm)	0.15
Matrix size	256 × 128
BW (kHz)	50
NA	1
AF	4
Acq. Time	1 min 36 s

Table A.2   Fig. 4.3

Pulse sequence	3D RARE
TE/TR (ms)	12/1500
FOV (mm)	5.2 × 2.3 × 2.3
Matrix size	200 × 88 × 88
BW (kHz)	50
NA	8
AF	8
Acq. Time	3 h 13 min 36 s

Table A.3   Fig. 4.5

Pulse sequence	3D FLASH
TE/TR (ms)	2.4/150
FOV (mm)	5.1 × 2.2 × 2.2
Matrix size	200 × 88 × 88
BW (kHz)	200
NA	6
Flip angle	40°
Acq. Time	1 h 56 min 9 s

Table A.4   Fig. 4.7

Pulse sequence	3D EPI
TE/TR (ms)	24/1000
FOV (mm)	2.3 × 8 × 4.4
Matrix size	118 × 400 × 220
BW (kHz)	100
NA	1
Nseg	16
Acq. Time	58 min 40 s

Table A.5   Fig. 4.9a

Pulse sequence	2D RARE
TE/TR (ms)	15/3000
FOV (mm)	12 × 8
thk (mm)	0.25
Matrix size	380 × 256
BW (kHz)	50
NA	1
AF	4
Acq. Time	12 min 48 s

Table A.6   Fig. 4.9b

Pulse sequence	2D MDEFT
TE/TR/TI (ms)	2.8/9.9/1050
FOV (mm)	12 × 8
thk (mm)	0.25
Matrix size	380 × 256
BW (kHz)	100
NA	16
Nseg	4
Flip angle	20°
Acq. Time	34 min 8 s

Table A.7 Figs. 4.11, 4.12

Pulse sequence	2D DW-SE
TE/TR (ms)	18/2500
FOV (mm)	12.3 × 11
thk (mm)	0.13
Matrix size	240 × 220
δ/Δ (ms)	3/10
BW (kHz)	50
NA	4
Acq. Time	36 min 40 s

Table A.8 Fig. 4.14

Pulse sequence	3D DP-FISP
TE/TR(ms)	2.6/5.2
FOV (mm)	4.7 × 0.8 × 0.6
Matrix size	190 × 32 × 24
δ/Δ (ms)	2.5/10
BW (kHz)	100
NA	7
Nseg	4
Flip angle	20°
Acq. Time	9 min 36 s

Table A.9 Fig. 5.1

Pulse sequence	2D MSME
TE/TR (ms)	10/1500
FOV (mm)	10 × 3.2
thk (mm)	0.15
Matrix size	200 × 64
BW (kHz)	50
NA	4
Acq. Time	6 min 24s

Table A.10 Fig. 5.2

Pulse sequence	2D FLASH
TE/TR (ms)	(a) 2.5/1500 (b) 10/1500
FOV (mm)	10 × 3.2
thk (mm)	0.15
Matrix size	200 × 64
BW (kHz)	150
Flip angle	30°
NA	4
Acq. Time	6 min 24 s

Table A.11 Fig. 5.3

Pulse sequence	2D RARE
TE/TR (ms)	11.6/2500
FOV (mm)	5 × 5
thk (mm)	2.5
Matrix size	128 × 128
BW (kHz)	100
NA	4
Acq. Time	5 min 20s

Table A.12 Fig. 5.4

Pulse sequence	2D MSME
TE/TR (ms)	10/1000
FOV (mm)	6 × 5
thk (mm)	0.15
Matrix size	128 × 108
BW (kHz)	50
NA	1
Acq. Time	1 min 48 s

Table A.13   Fig. 5.5

Pulse sequence	2D MSME
TE/TR (ms)	10/1000
FOV (mm)	(a) 6 × 1.1
	(b) 6 × 2.2
thk (mm)	0.15
Matrix size	200 × 64
BW (kHz)	50
NA	1
Acq. Time	1 min 4 s

Table A.14   Fig. 5.6

Pulse sequence	2D MSME
TE/TR (ms)	8.7/1000
FOV (mm)	9.7 × 2.2
thk (mm)	2.5
Matrix size	128 × 64
BW (kHz)	100
NA	1
Flip angle	(a) 113° / 227°
	(b) 90° / 180°
Acq. Time	1 min 1 s

Table A.15   Fig. 5.7

Pulse sequence	2D RARE
TE/TR (ms)	7.3/2000
FOV (mm)	2.2 × 2.2
thk (mm)	8
Matrix size	32 × 32
BW (kHz)	50
NA	1
AF	1
RG	(a) 203
	(b) 90.5
Acq. Time	1 min 4 s

Table A.16   Fig. 5.8

Pulse sequence	2D MSME
TE/TR (ms)	10/1000
FOV (mm)	5 × 6
thk (mm)	2
Matrix size	(a) 64 × 128
	(b) 128 × 128
BW (kHz)	50
NA	1
Acq. Time	2 min 8s

Table A.17   Fig. 5.9

Pulse sequence	2D FLASH
TE/TR (ms)	4/500
FOV (mm)	10 × 6
thk (mm)	4
Matrix size	256 × 128
BW (kHz)	400
Flip angle	30°
NA	1
Acq. Time	1 min 4 s

Table A.18   Fig. 5.10

Pulse sequence	2D EPI-STE
TE/TR (ms)	53/1500
FOV (mm)	25 × 25
thk (mm)	1
Matrix size	164 × 164
BW (kHz)	400
NA	64
Nseg	1
Acq. Time	1 min 36 s

References

Abe, Y., Nguyen, K. V., Tsurugizawa, T., Ciobanu, L., and Le Bihan, D. (2017). Unpublished results.

Abragam, A. (1961). *Principles of Nuclear Magnetism*. Clarendon Press, Oxford.

Aguayo, J. B., Blackband, S. J., Schoeniger, J., Mattingly, M. A., and Hintermann, M. (1986). Nuclear magnetic resonance imaging of a single cell, *Nature*, **322**, pp. 190–191.

Aschner, M., and Gannon, M. (1994). Manganese (Mn) transport across the rat blood-brain barrier: saturable and transferrin-dependent transport mechanisms, *Brain Res. Buli.*, **33**, pp. 345–349.

Badea, A., and Johnson, G. A. (2013). Magnetic resonance microscopy, *Stud. Health Technol. Inform.*, **185**, pp. 153–184.

Bennett, K. M., Schmainda, K. M., Bennett (Tong), R., Rowe, D. B., Lu, H., and Hyde, J. S. (2003). Characterization of continuously distributed cortical water diffusion rates with a stretched-exponential model, *Magn. Reson. Med.*, **50**, pp. 727–734.

Bowtell, R. W., Brown, G. D., Glover, P. M., McJury, M., and Mansfield, P. (1990). Resolution of cellular structures by NMR microscopy at 11.7 T, *Philos. Trans. R. Soc. Lond. A*, **333**, pp. 457–467.

Brembs, B., Lorenzetti, F. D., Reyes, F. D., Baxter, D. A., and Byrne, J. H. (2002). Operant reward learning in *Aplysia*: neuronal correlates and mechanisms, *Science*, **296**, pp. 1706–1709.

Brurok, H., Schjitt, J., Berg, K., Karlsson, J. O., and Jynge, P. (1997). Manganese and the heart: acute cardiodepression and myocardial accumulation of manganese, *Act. Physiol. Scand.*, **159**, pp. 33–40.

Byerly, S. C., Chase, P. B., and Stimers, J. R. (1985). Permeation and interaction of divalent cations in calcium channels of snail neurons, *J. Gen. Physiol.*, **85**, pp. 491–518.

Cahill, L. S., Laliberté, C. L., Ellegood, J., Spring S., Gleave, J. A., Eede, M. C., Lerch, J. P., and Henkelman, R. M. (2012). Preparation of fixed mouse brains for MRI, *Neuroimage*, **60**, pp. 933–939.

Callaghan, P. T. (1991). *Principles of Nuclear Magnetic Resonance Microscopy*. Clarendon Press, Oxford.

Campbell, J. S., Siddiqi, K., Rymar, V. V., Sadikot, A. F., and Pike, G. B. (2005). Flow-based fiber tracking with diffusion tensor and q-ball data: Validation and comparison to principal diffusion direction techniques, *Neuroimage*, **27**, pp. 725–736.

Carr, H. Y., and Purcell, E. M. (1954). Effects of diffusion on free precession in nuclear magnetic resonance experiments, *Phys. Rev.*, **94**, pp. 630–638.

Cho, Z. H., Ahn, C. B., Juh, S. C., Jo, J. M., Friedenberg, R. M., Fraser, S. E., and Jacobs, R. E. (1990). Recent progress in NMR microscopy towards cellular imaging, *Philos. Trans. R. Soc. Lond. A*, **333**, pp. 469–475.

Cho, Z. H., Ahn, C. B., Juh, S. C., Lee, K. H., Jacobs, R. E., Lee, S., Yi, J. H., and Jo, J. M. (1988). Nuclear magnetic resonance microscopy with 4-microns resolution: theoretical study and experimental results, *Med. Phys.*, **15**, pp. 815–824.

Ciobanu, L., and Pennington, C. H. (2003). 3D micron-scale MRI of single biological cells, *Solid State NMR*, **25**, pp. 138–141.

Ciobanu, L., Seeber, D., and Pennington, C. H. (2002). 3D MR microscopy with resolution 3.7 microm by 3.3 microm by 3.3 microm, *J. Magn. Reson.* **158**,1–2, pp. 178–182.

Codd, S. L., and Seymour, J. S. (2009). *Magnetic Resonance Microscopy*. WILEY-VCH, Weinheim.

Crossgrove, J. S., and Yokel, R. A. (2005). Manganese distribution across the blood-brain barrier. IV. Evidence for brain influx through store-operated calcium channels, *Neurotoxicology*, **26**, pp. 297–307.

Dazai, J., Spring, S., Cahill, L. S., and Henkelman, R. M. (2011). Multiple-mouse neuroanatomical magnetic resonance imaging, *J. Vis. Exp.*, **48**, e2497.

Deoni, S. C., Peters, T. M., and Rutt, B. K. (2004). Quantitattive diffusion imaging with steady-state free precession, *Magn. Reson. Med.*, **51**, pp. 428–433.

Dhenain, M., Delatour, B., Walczak, C., and Volk, A. (2006). Passive staining: A novel ex vivo MRI protocol to detect amyloid deposits in mouse models of Alzheimer's disease, *Magn. Reson. Med.*, **55**, pp. 687–693.

Dietrich, O., Hubert, A., and Heiland, S. (2014) Imaging cell size and permeability in biological tissue using the diffusion-time dependence of the apparent diffusion coefficient, *Phys. Med. Biol.*, **59**, pp. 3081–3096.

Dodd, S. J., Williams, M., Suhan, J. P., Williams, D. S., Koretsky, A. P., and Ho, C. (1999). Detection of single mammalian cells by high-resolution magnetic resonance imaging, *Biophys. J.*, **76**, pp. 103–109.

Does, M. D., Parsons, E. C., and Gore, J. C. (2003). Oscillating gradient measurements of water diffusion in normal and globally ischemic rat brain, *Magn. Reson. Med.*, **49**, pp. 206–215.

Doneva, M., Börnert, P., Eggers, H., Stehning, C., Sénégas, J., and Mertins, A. (2010). Compressed sensing reconstruction for magnetic resonance parameter mapping. *Magn. Reson. Med.*, **64**, pp. 1114–1120.

Donoho, D. (2006). Compressed sensing, *IEEE Trans. Inf. Theory*, **52**, 4, pp. 1289–1306.

Einstein, A. (1905). On the movement of small particles suspended in stationary liquids required by the molecular-kinetic theory of heat (in German) *Ann. d. Phys.*, **17**, pp. 549–560.

Eroglu, S., Gimi, B., Roman, B., Friedman, G., and Magin, R. L. (2003). NMR spiral surface microcoils: field characteristics and applications. *Concepts Magn. Reson. Imag. Part B* **17B**, 1, pp. 1–10.

Flint, J., Hansen, B., Vestergaard-Poulsen, P., and Blackband, S. J. (2009). Diffusion weighted magnetic resonance imaging of neuronal activity in the hippocampal slice model, *Neuroimage*, **46**, pp. 411–418.

Flint, J. J., Lee, C. H., Hansen, B., Fey, M., Schmidig, D., Bui, J. D., King, M. A., Vestergaard-Poulsen, P., and Blackband, S. J. (2009). Magnetic resonance microscopy of mammalian neurons, *Neuroimage*, **46**, pp. 1037–1040.

Flint, J. J., Hansen, B., Portnoy, S., Lee, C. H., King, M. A., Fey, M., Vincent, F., Stanisz, G. J., Vestergaard-Poulsen, P., and Blackband, S. J. (2013). Magnetic resonance microscopy of human and porcine neurons and cellular processes, *Neuroimage*, **60**, pp. 1404–1411.

Fukushima, E., and Roeder, S. B. W. (1981). *Experimental Pulse NMR: A Nuts and Bolts Approach*, Addison-Wesley, New York.

Gbel, K., Gruschke, O. G., Leupold, J., Kern, J. S., Has, C., Bruckner-Tuderman, L., Hennig, J., von Elverfeldt, D., Baxan, N., and Korvink, J. G. (2015). Phased-array of microcoils allows MR microscopy of ex vivo human skin samples at 9.4 T, *Skin Res. Technol.*, **21**, 1, pp. 61–68.

Geiger, J. E., Hickey, C. M., and Magoski, N. S. (2009). Ca^{2+} entry through a non-selective cation channel in Aplysia bag cell neurons, *Neuroscience*, **162**, pp. 1023–1038.

Glover, P., and Mansfield, S. P. (2002). Limits to magnetic resonance microscopy, *Rep. Prog. Phys.*, **65**, pp. 1489–1511.

Grant, S. C., Buckley, D. L., Gibbs, S., Webb, A. G., and Blackband, S. J. (2001). MR microscopy of multicomponent diffusion in single neurons, *Magn. Reson. Med.*, **46**, pp. 1107–1112.

Griswold, M. A., Jakob, P. M., Heidemann, R. M., Nittka, M., Jellus, V., Wang, J., Kiefer, B., and Haase, A. (2002). Generalized autocalibrating partially parallel acquisitions (GRAPPA), *Magn. Reson. Med.*, **47**, 6, pp. 1202–1210.

Gruschke, O. G., Baxan, N., Clad, L., Kratt, K., von Elverfeldt, D., Peter, A., Hennig, J., Badilita, V., Wallrabe, U., and Korvink, J. G. (2012). Lab on a chip phased-array MR multi-platform analysis system, *Lab. Chip.* **12**, 3, pp. 495–502.

Haake, E. M., Brown, R. W., Thomson, M. R., and Venkatesam, R. (1999). *Magnetic Resonance Imaging.* John Wiley & Sons, New York.

Haase, A., Frahm, F., Matthaei, D., Hanicke, W., and Merboldt, K. D. (1986). FLASH imaging. Rapid NMR imaging using low flip-angle pulses, *J. Magn. Reson.*, **67**, 2, pp. 258–266.

Hargreavis, B. (2012). Rapid gradient echo imaging, *J. Magn. Reson. Imaging*, **36**, 6, pp. 1300–1313.

Henning, J. (1991). Echoes: how to generate, recognize, use or avoid them in MR-imaging sequences. Part I: Fundamental and not so fundamental properties of spin echoes, *Con. Magn. Reson.*, **3**, pp. 125–143.

Henning, J., Nauerth, A., and Friedburg, H. (1986). RARE imaging: A fast imaging method for clinical MR, *Magn. Reson. Med.*, **3**, 6, pp. 823–833.

Herberholz, J., Mims, C. J., Zhang, X., Hu, X., and Edwards, D. H. (2004). Anatomy of a live invertebrate revealed by manganese-enhanced magnetic resonance imaging, *J. Exp. Biol.*, **207**, pp. 4543–4550.

Herberholz, J., Mishra, S. H., Uma, D., Germann, M. W., Edwards, D. H., and Potter, K. (2011). Non-invasive imaging of neuroanatomical structures and neural activation with high-resolution MRI, *Front. Behav. Neurosci.*, **5**, pp. 1–9.

Hill, H. D. W., and Richard, R. E. (1968). Limits of measurement in magnetic resonance, *J. Phys. E: Sci. Instrum.* **1**, 10, pp. 977–983.

Hoult, D. I., and Richard, R. E. (1976). The signal-to-noise ratio of the nuclear magnetic resonance experiment, *J. Magn. Reson.* **24**, 1–2, pp. 71–85.

Hsu, E. W., Aiken, N. R., and Blackband, S. (1996). Nuclear magnetic resonance microscopy of single neurons under hypotonic perturbation, *Am. J. Physiol.*, **271**, pp. C1895–C1900.

Jackson, J. D. (1975). Classical Electrodynamics. Wiley, New York.

Jelescu, I. O., Ciobanu L., Geffroy, F., Marquet P., and Le Bihan, D. (2014). Effects of hypotonic stress and ouabain on the apparent diffusion coefficient of water at cellular and tissue levels in *Aplysia, NMR Biomed.*, **27**, pp. 280–290.

Jelescu, I. O., Nargeot, R., Le Bihan, D., and Ciobanu, L. (2013). Highlighting manganese dynamics in the nervous system of *Aplysia californica* using MEMRI at ultra-high field, *Neuroimage*, **76**, pp. 264–271.

Jensen, J. H., Helpern, J. A., Ramani, A., Lu, H., and Kaczynski, K. (2005). Diffusional kurtosis imaging: The quantification of non-gaussian water diffusion by means of magnetic resonance imaging, *Magn. Reson. Med.*, **53**, pp. 1432–1440.

Johnson, G. A., Thompson, M. B., Gewalt, S. L., and Hayes, C. E. (1986). Nuclear magnetic resonance imaging at microscopic resolution, *J. Magn. Reson.*, **68**, pp. 129–137.

Kandel, E. R. (1979). *Behavioral Biology of Aplysia. A Contribution to the Comparative Study of Opisthobranch Molluscs.* Freeman, San Francisco.

Kandel, E. R. (1983). From metapsychology to molecular biology: explorations into the nature of anxiety, *Am. J. Psychiatry*, **140**, pp. 277–293.

Kempsell, A. T., and Fieber, L. A. (2015). Aging in sensory and motor neurons results in learning failure in *Aplysia californica, PLOS ONE,* DOI:10.1371/journal.pone.0127056.

Koo, C., Godley, R. F., McDougall, M. P., Wright, S. M., and Han, A. (2014). A microfluidically cryocooled spiral microcoil with inductive coupling for MR microscopy, *IEEE Trans Biomed Eng.* **61**, 1, pp. 76–84.

Kupfermann, I. (1974). Feeding behavior in *Aplysia*: a simple system for the study of motivation, *Behav. Biol.*, **10**, pp. 1–26.

Lauterbur, P. C. (1973). Image formation by induced local inetractions: Examples employing nuclear magnetic resonance. *Nature*, **242**, pp. 190–191.

Le Bihan, D. (2014). Diffusion MRI: what water tells us about the brain, *EMBO Mol Med*, **6**, pp. 569–573.

Le Bihan, D. (1988). Intravoxel incoherent motion imaging using steady-state free precession, *Magn. Reson. Med.*, **7**, pp. 346–351.

Le Bihan, D. (2007). The 'wet mind': water and functional neuroimaging, *Phys. Med. Biol.*, **52**, pp. R57.

Le Bihan, D., Mangin, J. F., Poupon, C., Clark, C. A., Pappata, S., Molko, N., and Chabriat, H. (2001). Diffusion tensor imaging: concepts and applications, *J. Magn. Reson. Imaging*, **13**, pp. 534–546.

Le Bihan, D., Urayama, S., Aso, T., Hanakawa, T., and Fukuyama, H. (2006). Direct and fast detection of neuronal activation in the human brain with diffusion MRI, *Proc. Natl. Acad. Sci. USA*, **103**, pp. 8263–8268.

Lee, C. H., Flint, J. J., Hansen, B., and Blackband, S. J. (2015). Investigation of the subcellular architecture of L7 neurons of *Aplysia californica* using magnetic resonanc microscopy (MRM) at 7.8 microns, *Scientific Reports*, **5**, pp. 1–11.

Lee, S.-C., Kim, K., Kim, J., Lee, S., Yi, J., Kim, S., Ha, K., and Cheong, C. (2001). One micrometer resolution NMR microscopy, *J. Magn. Reson.*, **150**, pp. 207–213.

Lin, Y.-J., and Koretsky, A. P. (1997). Manganese ion enhances T1-weighted MRI during brain activation: An approach to direct imaging of brain function, *Magn. Reson. Med.*, **38**, pp. 378–388.

Liu, Y., Sajja, B. R., Gendelman, H. E., and Boska, M. D. (2013). Mouse brain fixation to preserve *in vivo* manganese enhancement for *ex vivo* MEMRI, *JMRI*, **38**, pp. 482–487.

Lustig, M., Donoho, D., and Pauly, J. M. (2007). Sparse MRI: the application of compressed sensing for rapid MR imaging, *Magn. Reson. Med.*, **58**, 6, pp. 1182–1195.

Mansfield, P. (1977). Multi-planar image formation using NMR spin echoes, *J. Phys. C.*, **105**, pp. L55–L58.

McClymont, D., Teh, I., Whittington, H. J., Grau, V., and Schneider, J. E. (2016). Prospective acceleration of diffusion tensor imaging with compressed sensing using adaptive dictionaries. *Magn. Reson. Med.*, **76**, pp. 248–258.

Medhurst, R. (1947). H. F. resistance and self-capacitance of single-layer solenoids, *Wireless Eng.*, **35**, pp. 80–92.

Meiboom, S., and Gill, D. (1954). Modified spin-echo method for measuring nuclear relaxation times, *Rev. Sci. Instrum.*, **29**, pp. 688.

Mena, I., Marin, O., Fuenzalida, S., and Cotzias, G. C. (1967). Chronic manganese poisoning: clinical picture and manganese turnover, *Neurology*, **17**, pp. 128–136.

Minard, K. R., and Wind, R. A. (2001). Solenoidal microcoil design. Part I: Optimizing RF homogeneity and coil dimensions, *Concepts Magn. Res* **13**, 2, pp. 128–142.

Minard, K. R., and Wind, R. A. (2001). Solenoidal microcoil design. Part II: Optimizing winding parameters for maximum signal-to-noise performance, *Concepts Magn. Res* **13**, 3, pp. 190–210.

Mitra, P. P., Sen, P. N., and Schwartz, L. M. (1993). Short-time behavior of the diffusion coefficient as a geometrical probe of porous media, *Phys. Rev. B*, **47**, pp. 8565–8574.

Moussavi-Biugui, A., Stieltjes, B., Fritzsche, K., Semmler, W., and Laun, F. B. (2011). Novel spherical phantoms for q-ball imaging under in vivo conditions, *Magn. Reson. Med.*, **65**, pp. 190–194.

Nabuurs, R. J. A., Hegeman, I., Natte, van Duinen, S. G., van Buchem, M. A. van der Weerd, A., and Webb, A. G. (2011). High-field MRI of single histological slices using an inductively coupled, self-resonant microcoil: application to ex vivo samples of patients with Alzheimer's disease. *NMR Biomed.* **24**, pp. 351–357.

Nargeot, R., Baxter, D. A., and Byrne, J. H. (1997). Contingent-dependent enhancement of rhythmic motor patterns: an in vitro analog of operant conditioning, *J. Neuros*, **17**, pp. 8093–8105.

Nargeot, R., Petrissans, C., and Simmers, J. (2007). Behavioral and in vitro correlates of compulsive-like foodseeking induced by operant conditioning in *Aplysia*, *J. Neurosci.*, **27**, pp. 8059–8070.

Narita, K., Kawasaki, F., and Kita, H. (1990). Mn and Mg influxes through Ca channels of motor nerve terminals are prevented by verapamil in frogs, *Brain Res*, **510**, pp. 289–295.

Nelson, M. T. (1986). Interactions of divalent cations with single calcium channels from rat brain synaptosomes, *J. Gen. Physiol.*, **87**, pp. 201–222.

Nguyen, K. V., Le Bihan, D., Ciobanu, L., and Li, J.-R. (2017). Unpublished results.

Nguyen, K. V., Li, J. R., Radecki, G., and Ciobanu, L. (2015). DLA based compressed sensing for high resolution MR microscopy of neuronal tissue, *J. Magn. Reson.*, **259**, pp. 186–189.

Ogawa, S., Lee, T. M., Ray, A. R., and Tank, D. W. (1990). Brain magnetic resonance imaging with contrast dependent on blood oxygenation, *Proc. Natl. Acad. Sci. U. S. A.*, **87**, pp. 9868–9872.

Olson, D. L., Peck, T. L., Webb, A. G., Magin, R. L., and Sweedler, J. V. (1995). High-resolution microcoil 1H-NMR for mass-limited, nanoliter-volume samples. *Science* **270**, 5244, pp. 1967–1970.

Pautler, R. G., and Koretsky, A. P. (2002). Tracing odor-induced activation in the olfactory bulbs of mice using manganese-enhanced magnetic resonance imaging, *Neuroimage*, **16**, pp. 441–448.

Peck, T. L., Magin, R. L., and Lauterbur, P. C. (1995). Design and analysis of microcoilsfor NMR microscopy, *J. Magn. Res. B* **108**, 2, pp. 114–124.

Pruessmann, K. P., Weiger, M., Scheidegger, M. B., and Boesiger, P. (1999). SENSE: sensitivity encoding for fast MRI, *Magn. Reson. Med.*, **42**, 5, pp. 952–962.

Pullens, P., Roebroeck, A., and Goebel, R. (2010). Ground truth hardware phantoms for validation of diffusion-weighted MRI applications, *J. Magn. Reson. Imag.*, **32**, pp. 482–488.

Radecki, G., Nargeot, R., Jelescu, I. O., Le Bihan, D., and Ciobanu, L. (2014). Functional magnetic resonance microscopy at single-cell resolution in *Aplysia californica*, *Proc. Natl. Acad. Sci. USA*, **111**, pp. 8667–8672.

Sattin, W., Mareci, T., and Scott, K. (1985). Exploiting the stimulated echo in nuclear magnetic resonance imaging, *J. Magn. Reson.*, **64**, pp. 177–182.

Schiavi, S., Haddar, H., and Li, J.-R. (2016). Correcting the short time ADC formula to account for finite pulses, ISMRM Workshop, Singapore, 2016.

Schipper, H. M. (2012). Neurodegeneration with brain iron accumulation: clinical syndromes and neuroimaging, *Biochim. Biophys. Acta.*, **1822**, pp. 350–360.

Schoeniger, J. S., Aiken, N., Hsu, E., and Blackband, S. J. (1994). Relaxation-time and diffusion NMR microscopy of single neurons, *J. Magn. Reson. Ser. B*, **103**, pp. 261–273.

Seeber, D. A., Hoftiezer, J. H., Daniel, W. B., Rutgers, M. A., and Pennington, C. H. (2000). Triaxial magnetic field gradient system for microcoil magnetic resonance imaging. *Rev. Sci. Instrum.*, **71**, 11, pp. 4263–4272.

Sehy, J. V., Ackerman, J. J. H., and Neil, J. J. (2001). Water and lipid MRI of *Xenopus* oocyte, *Magn. Reson. Med.*, **46**, pp. 900–907.

Sehy, J. V., Ackerman, J. J. H., and Neil, J. J. (2002). Evidence that both fast and slow water ADC components arise from intracellular space, *Magn. Reson. Med.*, **48**, pp. 765–770.

Sehy, J. V., Ackerman, J. J. H., and Neil, J. J. (2002). Apparent diffusion of water, ions and small molecules in teh Xenopus oocyte is consistent with Brownian displacement, *Magn. Reson. Med.*, **48**, pp. 42–51.

Sehy, J. V., Banks, A. A., Ackerman, J. J. H., and Neil, J. J. (2002). Importance of intracellular water apparent diffusion to the measurement of membrane permeability, *Biophys. J.*, **83**, pp. 2856–2863.

Shepherd, T. M., Blackband, S. J., and Wirth, E. D. (2002). Simultaneous diffusion MRI measurements from multiple perfused rat hippocampal slices, *Magn. Reson. Med.*, **48**, pp. 565–569.

Shepherd, T. M., Thelwall, P. E., Stanisz, G. J., and Blackband, S. J. (2009). Aldehyde fixative solutions alter the water relaxation and diffusion properties of nervous tissue, *Magn. Reson. Med.*, **62**, pp. 26–34.

Silva, A. C., Lee, J. H., Aoki, I., and Koretsky, A. P. (2004). Manganese-enhanced magnetic resonance imaging (MEMRI): Methodological and practical considerations, *NMR Biomed.*, **17**, pp. 532–543.

Sun, S. W., Neil, J. J., Liang, H. F., He, Y. Y., Schmidt, R. E., Hsu, C. Y., and Song, S.-K. (2005). Formalin fixation alters water diffusion coefficient magnitude but not anisotropy in infarcted brain, *Magn. Reson. Med.*, **53**, pp. 1447–1451.

Svehla, P., Nargeot, R., and Ciobanu, L. (2017). Unpublished results.

Tang, T. A., and Jerschow, A. (2010). Practical aspects of liquid-state NMR with inductively coupled solenoid coils. *Magn. Reson. Chem.* **48**, pp. 63–770.

Tannus, A., and Garwood, M. (1997). Adiabatic pulses, *NMR Biomed.*, **10**, pp. 423–434.

Tsurugizawa, T., Ciobanu, L., and Le Bihan, D. (2013). Water diffusion in brain cortex closely tracks underlying neuronal activity, *Proc. Natl. Acad. Sci. USA*, **110**, pp. 11636–11641.

Van der Linden, A., Van Meir, V., Tindemans, I., Verhoye, M., and Balthazart, J. (2004). Applications of manganese-enhanced magnetic resonance imaging (MEMRI) to image brain plasticity in song birds. *NMR Biomed.*, **17**, pp. 602–612.

van Duijn, S., Nabuurs, R. J. A., van Rooden, S., Maat-Schieman, M. L. C., van Duinen, S. G., van Buchem, M. A., van der Weerd, L., and Natté, R. (2011). MRI artifacts in human brain tissue after prolonged formalin storage, *Magn. Reson. Med.*, **65**, pp. 1750–1758.

Wang, T., Ciobanu, L., Zhang, X. Z., and Webb, A. G. (2008). Inductively coupled RF coil design for simultaneous microimaging of multiple samples, *Concepts Magn. Reson.* **33B**, pp. 236–243.

Webster, M., Witkin, K. L., and Cohen-Fix, O. (2009). Sizing up the nucleus: nuclear shape, size and nuclear-envelope assembly, *J. Cell Sci.*, **122**, pp. 1477–1486.

Webb, A. (2010). Microcoils, *eMagRes* **1**, pp. 1–6.

Weiger, M., Schmidig, D., Denoth, S., Massin, C., Vincent, F., Schenkel, M., and Fey, M. (2008). NMR-microscopy with isotropic resolution of 3.0 μm using dedicated hardware and optimized methods, *Concepts Magn. Reson. Part B*, **33B**, pp. 84–93.

Wu, Z., Mittal, S., Kish, K., Yu, Y., Hu, J., and Haake, E. M. (2009). Identification of calcification with magnetic resonance imaging using susceptibility-weighted imaging: a case study, *J. Magn. Reson. Imaging*, **29**, pp. 177–182.

Yablonskiy, D. A., Bretthorst, G. L., and Ackerman, J. J. H. (2003). Statistical Model for Diffusion Attenuated MR Signal, *Magn. Reson. Med.*, **50**, pp. 664–669.

Zhang, Z., Seginer, A., and Frydman, L. (2016). Single-scan MRI with exceptional resilience to field heterogeneities, *Magn. Reson. Med*, **77**, pp. 623–634.

Zhou, X., Potter, C. S., Lauterbur, P. C., and Voth, B. (1989). 3D microscopic NMR imaging with $(6.37 \text{ micron})^3$ isotropic resolution, *Eighth Annu. Meet. Soc. Magn. Reson. Med.*, Amsterdam.

Index